HOW TO FLY A SECOND WORLD WAR HEAVY BOMBER

LANCASTER HALIFAX STIRLING

Ed. L. Archard

AMBERLEY

All images are courtesy of Campbell McCutcheon unless otherwise stated.

First published 2014

Amberley Publishing
The Hill, Stroud
Gloucestershire, GL5 4EP

www.amberley-books.com

British Library Cataloguing in Publication Data.
A catalogue record for this book is available from the British Library.

ISBN 978 1 4456 3672 6 (print)
ISBN 978 1 4456 3690 0 (ebook)

Typeset in 10pt on 12pt Sabon.
Typesetting and Origination by Amberley Publishing.
Printed in the UK.

CONTENTS

Introduction

A crowd of young men sit in a hall, looking towards a stage with a thick curtain drawn across, covering the back wall. A senior officer comes on to the stage, sweeps the curtain back and announces, 'The target for tonight is …' After the briefing, the crews clamber onto trucks and into buses to be taken to their aircraft scattered around the airfield: huge, four-engined heavy bombers, painted black. This is the stereotypical image of RAF Bomber Command, courtesy of Air Ministry propaganda films from during the war, like *Target for Tonight*, and of the many films produced after the war, perhaps most famously *The Dambusters*. In *The Dambusters*, of course, the aircraft that Guy Gibson (as played by Richard Todd) and his comrades boarded for their mission were Avro Lancasters. The Lancaster was the backbone of Bomber Command and by the end of the war some three quarters of Bomber Command squadrons flew the Lancaster. However, there were two other four-engined heavy bombers used by the RAF during the war: the Short Stirling and the Handley Page Halifax.

When the Second World War broke out, none of these big, four-engined bombers were in service with the RAF, only medium bombers like the Vickers Wellington, the Handley Page Hampden and the Armstrong Whitworth Whitley; only the Wellington would prove to be able to survive the German defences. The first bombing raids carried out against Germany were raids against warships at Wilhelmshaven, and at Brunsbüttel on the Kiel Canal. A fear of inflicting civilian casualties led to Bomber Command concentrating on dropping propaganda leaflets thereafter, and the leaflet campaign lasted into 1940 before it was cancelled. During and after the Fall of France, Bomber Command was authorised to attack industrial targets in Germany linked to the German war effort, and also to attack the concentrations of barges in the Channel ports intended to carry the German invasion force. It was at this point in the war that the first of our heavy bombers came into service.

Short Stirling

The Short Stirling, the first of the RAF's heavy bombers, made its maiden flight on 14 May 1939, but did not fly its first combat mission until February 1941. It had been designed in response to Air Ministry Specification B.12/36, which had been looking for something very different: an aircraft to deal with emergencies in the furthest-flung corners of the British Empire, and which could transport twenty-four troops and then carry out bombing missions to support them when they went into action.

Short Brothers, then based at Rochester in Kent, were best known for their flying boats; flying boats were widely used in civil aviation in the 1920s and 1930s because airfields suitable for large aircraft weren't common and a flying boat could land on any stretch of calm water. Although Short Brothers were not at first asked to submit a design for B.12/36, their experience with large, long-range aircraft and the designs for flying boats of the right size which they had in hand at the time meant that they were soon invited to submit a proposal. The initial Short's design, the S.29, was based on their Sunderland flying boat; more or less all the designers had to do was remove the lower deck and the hull from the original flying boat design. The rugged Sunderland would go on to make a reputation for itself during the Second World War, flying long patrols over the North Atlantic looking for the U-boats that attacked the essential convoys from North America and occasionally landing to rescue survivors from torpedoed ships. Before its cousin could fly, however, the Air Ministry specified changes to the design, increasing its maximum altitude, reducing the wing span and also reducing its take-off distance.

After the final design for the Stirling was flown in May 1939, service production began at Rochester in August 1940, at the height of the Battle of Britain. The area was home to several famous aircraft firms and as a result was heavily bombed by the Luftwaffe; Stirling production was spread out, with orders going to factories in Belfast and just south of Birmingham.

The Stirling that went into production was powered by four Bristol Hercules XI radial engines (the Stirling was the only one of the RAF's three heavy bombers to be designed from the start to carry four engines – the others both started as twin-engine designs). As well as a 14,000 lb bombload (far heavier than any other British bomber at that time), the Stirling was fitted with nose and tail turrets armed with .303 Browning machine guns. There was also a retractable turret fitted in the belly of the aircraft, behind the bomb bay, but this was problematic because the turret had a tendency to drop down unexpectedly and hit the ground when the aircraft was taxiing over bumps. It was also cramped, and the escape hatch was nearly impossible to use. Including the three gunners (the front gunner doubled up as the radio operator), a Stirling carried seven crew: a first and second pilot, flight engineer and bomb aimer/navigator made up the numbers.

As well as the retractable belly turret, there were other problems with the Stirling. It was an ungainly machine and the limited wingspan reduced its maximum altitude dramatically. This was a problem when Stirlings were used to attack targets in Italy: instead of flying over the Alps, the bombers had to fly through them instead. The altitude proved to be an even more serious problem when the Halifax and Lancaster, both of which flew at greater height, came into service: Stirling pilots would often find themselves flying through bombs being dropped by the aircraft above them.

The first operational mission by a Stirling was on the night of 10/11 February 1941, when three Stirlings of No. 7 Squadron attacked fuel tanks near Rotterdam; after that they started to be used in greater and greater numbers until Halifaxes and Lancasters became increasingly available. Stirlings attacked targets in France such as the ports of Boulogne and Brest; after Area Bombing was put into practice, Stirlings attacked targets across Germany including Essen, Lübeck, Hamburg, Mannheim, Warnemünde and the first Thousand Bomber Raid against Cologne on 30 May 1942, as well as returning to France to attack the Renault works at Billancourt, outside Paris.

From December 1943, Stirlings started to be allocated away from the main bombing campaign against Germany, although they would be drafted in to make up the numbers for the Thousand Bomber Raids, an important part of the Air Ministry's propaganda campaign. As well as the performance issues, the design of the Stirling's bomb bay meant that it didn't have the space to carry the new 4,000-lb bombs that were coming into service. Instead, Stirling squadrons were deployed to other theatres of the war, in the Mediterranean, for instance.

In total, Stirlings flew 14,500 sorties for Bomber Command, dropping 27,000 tons of bombs. In the process, 582 Stirlings were lost in action with another 119 written off.

As well as bombing missions, the Stirlings were also used to tow heavily laden gliders for the airborne forces, and to drop paratroopers, doing both jobs on D-Day and as part of Operation Market Garden at Arnhem later in 1944. They were also used for so-called Gardening missions, laying mines in the shipping lanes outside ports in Germany and occupied Europe, and for dropping agents and supplies for the Special Operations Executive (SOE), the agents jumping into the darkness from an exit where the retractable belly turret had been.

Handley Page Halifax

The second of the RAF's heavy bombers, the Handley Page Halifax first flew on 25 October 1939. Handley Page, founded in 1909 and Britain's first publicly traded aircraft company, had designed and built a series of bombers for what would become the RAF during the First World War. The H.P.56, which would later become the Halifax, had been designed as a twin-engine aircraft in response to an Air Ministry specification, P.13/26, for a bomber that could be used around the world. Although the RAF were interested in four-engined bombers, there were design questions that were still being resolved by the Royal Aircraft Establishment at Farnborough and the H.P.56 had been designed with the idea of using twin Rolls-Royce Vulture engines. However, the Vulture was proving to be problematic, and Handley Page was asked to redesign the aircraft using different engines.

After the design was approved by the RAF, the Halifax went into production at Handley Page's factory in Lancashire, where more than 2,000 would be produced during the course of the war. Because of the promise of the design, the first 100 Mk 1 Halifaxes were ordered straight off the drawing board. The two Vultures had been replaced by four Rolls-Royce Merlin X engines. A Halifax Mark 1 could carry 13,000 lb of bombs in its bomb bay and wings, although like the Stirling it would face criticism because the structure of the bomb bay meant it could not carry the 4,000 lb bombs which were an integral part of Sir Arthur Harris's strategy for bombing German towns and cities. To defend itself against German night fighters, the Halifax carried .303 machine guns in turrets in the nose and tail, and in some cases in the waist of the aircraft as well. In addition to the pilot and the two gunners, the Halifax carried a flight engineer, navigator, bomb aimer and radio operator/gunner for a total of seven crew. The flight engineer replaced the co-pilot of earlier bombers because it was not an effective use of a trained pilot; instead, the flight engineer helped the pilot by monitoring the myriad instruments in the cockpit.

Halifaxes first entered service with the RAF in November 1940 and first saw combat in a mission on the night of 11/12 March 1941, in a raid against the port of Le Havre in northern France. In July 1941, Halifax bombers would attack the German pocket battleship *Scharnhorst* in the French port of La Pallice, and later in the year the Halifaxes returned to the French coast to mount another attack on *Scharnhorst* and her sister *Gneisnau*, which were considered to be a serious threat to the Atlantic convoys. After the Stirling was retired from Bomber Command's offensive against Germany, the Halifaxes took their place in the bottom layer of the bomber stream. This was the most dangerous position because of the flak, but also because of the night fighters. In early 1944, Messerschmitt 110s were replaced in the German night fighter force by Junkers Ju88 bombers modified to carry on-board radar and cannon mounted to fire above and slightly forward of the night fighter; this was known as *Schräge Musik* (literally 'slanted music' – the German term for jazz) and the idea was that the night fighter would be able to fire up into the belly of a bomber, laden with explosives, incendiaries and fuel. In their new position, the Halifax crews were particularly vulnerable to these new tactics and suffered accordingly: in January 1944, more than 10 per cent of No. 4 Group's Halifax aircraft failed to return from Germany, a casualty rate that led to the Halifaxes too being moved to other missions.

Like the Stirling, the Halifax was used to tow gliders and for special operations, dropping agents and supplies to resistance movements in occupied Europe. It was also used for anti-submarine patrols by Coastal Command (a role in which Sir Arthur Harris had been very reluctant to use his long-range bombers). Most interestingly, Halifaxes were used for electronic warfare by Bomber Command's 100 Group, a force set up to counter the Germans' lethal radar-equipped night fighters. In August 1942, for instance, Halifaxes had also flown some of the first missions for Bomber Command's Pathfinder Force (set up to identify and mark targets for the main bomber force, to counter problems in navigation and bomb aiming generally) against Flensburg. Halifaxes equipped with the H2S blind bombing system (which used a radar transmitter mounted in the aircraft to scan the ground) also flew in the raid against Hamburg.

Avro Lancaster

The Avro Lancaster was the third of the RAF's four-engined heavy bombers, and is still by far the most famous. The Lancaster was the descendant of the Avro Manchester, which had been designed in response to the same Air Ministry Specification that had produced the Handley Page Halifax, P.13/26; the Manchester had also been intended to be a twin-engine aircraft, powered by two Rolls-Royce Vultures, but the Vulture proved to be chronically unreliable and was the death of the Manchester. Avro's chief designer, Roy Chadwick, produced a new version of the Manchester powered by four of Rolls-Royce's Merlin engines, which were becoming just as powerful as the Vulture had been intended to be and were considerably more reliable; this became the Lancaster.

Avro's new design first flew from Ringway Airport, Manchester, where the company had a testing facility, and it was almost immediately recognised as a superb design. Over 7,300 Lancasters were built over the course of the war, more than 400 of them by Victory Aircraft in Ontario, Canada. The Lancaster was powered by four Merlin XX engines and, like the other two heavy bombers, carried a seven-man crew and had nose and rear turrets for .303 machine guns, although the Lancaster had a turret on top of the fuselage as well. Later models of the Lancaster would carry Bristol Hercules engines (the BII) and Merlin engines built by Packard in the United States. Lancasters would also be widely fitted with electronic devices to aid navigation, to aid bombing accuracy in bad weather conditions and to counter the radar used by the German early warning systems and night fighters. As well as the H2S system, Lancasters would carry Gee (a navigation system that picked up radio pulses from the UK), Monica (a rear-looking radar system to warn of night fighters), Boozer (which was designed to warn a crew if their aircraft was being tracked by German radar) and Airborne Cigar (a system only carried by 101 Squadron that allowed a German-speaker to feed false information to the German night fighters and jam their radio). Lancasters would also carry Window, paper strips treated with aluminium that could be dropped in large quantities to create false signals on a radar display.

No. 44 Squadron was the first RAF unit to convert to the Lancaster, receiving its first aircraft in late 1941 and flying its first operational mission on 3 March 1942. From then on the aircraft was the backbone of Bomber Command, equipping three-quarters of its squadrons. In all, between 1942 and 1945 Lancasters flew 156,000 sorties and dropped more than 600,000 tons of bombs. More than 3,000 Lancasters were lost in action, and only thirty-five completed more than 100 missions. Most of these sorties would have been area bombing raids, one of the most controversial strategies of the war: the design of the Lancaster's bomb bay meant that, unlike the other two heavy bombers, it could carry the 4,000-lb bombs which were intended to cause blast damage over a wide area and increase the effect of the incendiary bombs that were also carried. The Lancasters could also be adapted to carry the heaviest bombs designed for the RAF, and also the most unusual. Lancasters of 617 Squadron, of course, flew the Dams raid in May 1943 using Barnes Wallis's Upkeep bouncing bombs; 617 would also use 12,000-lb Tallboy bombs and 22,000-lb Grand Slam bombs to carry out precision attacks against targets such as the Bielefeld Viaduct and the battleship *Tirpitz*.

In late April and May 1945, the Lancasters of Bomber Command were used to drop food to civilians in the areas of the Netherlands that were still under German occupation in what was known as Operation Manna. As well as fighting the war in Europe, Lancasters were intended to make up the backbone of Tiger Force, a planned Commonwealth bomber force which was supposed to take part in the invasion of Japan, flying from bases on the island of Okinawa, but Japan surrendered before the invasion could take place. After the war, Lancasters were sold to Argentina and saw service in military coups there; more prosaically, Lancasters were used by the French navy in the 1950s and 1960s, and by the Canadians as well. Converted to transport aircraft, Lancasters saw service in the Berlin Airlift and in the development of in-flight refuelling, the civilian version being known as the Lancastrian. They would also serve as the ancestor of Avro's Lincoln bomber and York transport aircraft.

Which Was Best?

Inevitably, when a question like this comes up, many people will simply name the Lancaster, because of its fame. Halifax and Stirling crews would have argued the point, however. It is certainly true that the Lancaster carried out the most spectacular raids of the war, the Dams raid and the attack on the *Tirpitz* in particular. Some of Bomber Command's most famous pilots were Lancaster pilots: Guy Gibson, Leonard Cheshire and James Tait, who led the attack on the *Tirpitz*. The Lancaster was the most reliable of the three bombers but on the other hand, fewer crew members on average escaped from Lancasters than from the Halifax and the Stirling – Halifax crews seem to have had the best odds as the escape hatches were easily accessible. However, you might have had a worse flight in a Halifax: crews complained about the heating system, which seemed to be non-existent, which could be a major problem when flights lasted six or seven hours or maybe more, depending on the target, especially in winter. Although Stirling crews ran the risk of being bombed by their own comrades, thousands of feet above them in the bomber stream, the Stirling could out-turn the German night fighters, a definite advantage in the battle for survival.

All three of these aircraft would have seemed like something special in the early years of the Second World War. They were bigger and more sophisticated than anything that Bomber Command had ever used before, and to their crews they gave off a palpable sense of menace that inspired confidence, particularly in the black paint that was used to camouflage the aircraft for night flying. These aircraft were also one of the few ways in which British civilians could feel the country was hitting back against Germany and were therefore an important propaganda weapon. Although flying a bomber was not as exhilarating as flying a single-engine fighter like a Spitfire or a Hurricane, it nonetheless took a lot of skill to bring one back from a combat mission, particularly in one piece. Judge for yourself which is best.

Ground crew loading a Short Stirling with bombs for that night's mission.

Handley Page Halifax bombers of the Royal Australian Air Force lined up on an airfield somewhere in North Africa. When the Lancaster began to come into service in large numbers, Halifaxes and Stirlings were transferred away from raids against Germany.

Bomber Command in action. This artist's impression shows an air raid in progress against Hanover.

Glider troops loading a 6-pounder gun onto a Horsa glider with a Halifax towing aircraft in the background.

The Halifax doing what it was first designed for: an artist's impression of a raid over Kassel.

A Halifax carrying a jeep fitted with special shock absorbers to allow it to be dropped by parachute to airborne troops.

An Avro Lancaster on display in Trafalgar Square for the start of Wings for Victory Week on 6 March 1943.

Another view of a Lancaster following a mission, with the ground crew checking it over and making repairs.

A 4,000-lb 'Cookie' is loaded into the bomb bay of a Lancaster. The structure of the Lancaster's bomb bay meant it could carry bombs of this size and larger, but the Stirling and Halifax could not.

A 12,000-lb 'Tallboy' bomb being towed towards a waiting Lancaster. These bombs were carried on precision raids against targets such as the U-boat pens on the French Atlantic coast and V1 launch sites.

A 22,000-lb Grand Slam. The heaviest bomb carried by the RAF during the Second World War, it was designed, like the Tallboy, by Barnes Wallis and was used to sink the battleship *Tirpitz* at her anchorage in northern Norway.

SHORT STIRLING

STIRLING I. III & IV

NOTES TO USERS

THIS publication is divided into six parts: Descriptive, Handling, Operating Data, Emergencies, Data and Instructions for Flight Engineers and Illustrations. Part I gives only a brief description of the controls with which the pilot should be acquainted.

These Notes are complementary to A.P. 2095 Pilot's Notes General and assume a thorough knowledge of its contents. All pilots and flight engineers should be in possession of a copy of A.P. 2095 (*see* A.M.O.A93/43). Flight engineers should also have a copy of A.P. 2764 as soon as possible.

Words in capital letters indicate the actual markings on the controls concerned.

Additional copies may be obtained from A.P.F.S., Fulham Road, S.W.3, by application on R.A.F. Form 294A, in duplicate, quoting the number of this publication in full— A.P. 1660A, C & D—P.N.

Comments and suggestions should be forwarded through the usual channels to the Air Ministry (D.T.F.).

AIR MINISTRY
January 1944

AIR PUBLICATION 1660A, C & D—P.N.
Pilot's and Flight Engineer's Notes

STIRLING I, III and IV
PILOT'S & FLIGHT ENGINEER'S NOTES

3rd Edition. This edition supersedes all previous issues.

LIST OF CONTENTS

PART I—DESCRIPTIVE

PART IV—EMERGENCIES

PART V—DATA AND INSTRUCTIONS FOR FLIGHT ENGINEERS

PART VI—ILLUSTRATIONS

AIR PUBLICATION 1660A C & D—P.N.
Pilot's and Flight Engineer's Notes

PART I

DESCRIPTIVE

NOTE.—The numbers quoted in brackets after items in the text refer to the key numbers of the illustrations in Part VI.

FUEL AND OIL SYSTEMS

1. **Fuel Tanks.**—Seven tanks are fitted in each wing, and three auxiliary tanks may be carried in the wing bomb cells on each side. Capacities are:

		Each side	*Total*
Outer engine:	No. 3	63 galls.	
	No. 4	254 ,,	
	No. 5	164 ,,	
	No. 6	81 ,,	
		562 galls.	1,124 galls.
Inner engine:	No. 1	80 galls.	
	No. 2	331 ,,	
	No. 7	154 ,,	
		565 galls.	1,130 galls.
	Total normal ..	..	2,254 galls.
Auxiliary (Wing bomb cell) ..		219 galls.	438 galls.
Total (all tanks)	.. ..	..	2,692 galls.

All tanks with the exception of Nos. 7 are self-sealing. The latter should not be used unless maximum range is essential; they should never be used as well as the auxiliary wing bomb cell tanks unless large oil tanks (Mod. 601) are fitted. The contents of Nos. 2 and 4 tanks are jettisonable. Later aircraft have a nitrogen fire protection system fitted; the master valve and pressure gauge are on the front spar frame on the plate carrying the starboard heating system controls. This valve should be on at all times when the engines are running.

PART I—DESCRIPTIVE

2. **Fuel Cocks.**—The pilot controls four master fuel cocks (77) (labelled CARBURETTOR COCKS) in the cockpit roof. The flight engineer controls individual tank cocks (89, 91) at his station; the inter-engine balance cocks below the drag sheeting forward of the rear spar frame; and the inter-system balance cock under a hinged cover on the floor in front of the rear spar frame.

3. **Fuel gauges and warning lights**

(i) Contents gauges, with pushbuttons for each tank, are on the engineer's instrument panel. When taking a reading, particularly during refuelling, press the button until slight resistance is felt to release the float in the tank unit; wait for a second or two for the float to take up its proper level, and then push the button fully in to complete the gauge circuit. The gauge readings are approximate only but are within about ± 3 gallons.

(ii) Fuel pressure gauges are on the instrument panel. When these indicate falling pressure, select a fresh tank immediately to avoid engine cutting. The lights may flicker during take-off due to fuel surge; this does not necessarily indicate failing pressure.

4. **Oil System.**—Each engine oil tank holds 25½ (later aircraft 32) gallons, with 5½ gallons air space. Oil pressure and temperature gauges and oil dilution pushbuttons are at the flight engineer's station. Oil is circulated in the usual manner, the oil cooler being fitted with two relief valves; the carburettors are oil heated.

MAIN SERVICES

5. **Electrical system.**—Early aircraft have a 24-volt, 1,000-watt generator on each inboard engine, connected in parallel with two 40-amp. hr. batteries. Later aircraft have a 24-volt, 1,500-watt generator on each inboard engine, connected with four 40-amp. hr. batteries. A GROUND-FLIGHT switch on the rear spar frame isolates the batteries when the aircraft is parked and when using a ground starter battery. A plug-in socket for the external supply is near the switch.

PART I—DESCRIPTIVE

Services: Undercarriage and tail wheels
Flaps
Bomb doors
Bomb release gear
Pressure-head heating
Radio
D.R. Compass
Navigation lights
Glove and boot heating
Engine starters
Cowling gills
Internal lighting
Propeller feathering motors
Mark XIV bombsight

6. **Hydraulic system.**—The port inner engine is fitted with a twin pump for operating the nose and mid-upper turrets. The starboard inner engine is fitted with a pump for operating the tail turret. On later aircraft there is an additional pump, for the mid under turret, driven by the port inner engine.

7. **Pneumatic systems**

(i) *Vacuum system:*

(*a*) Pumps, one on each inner engine, supply the instruments and the Mk. XIV bombsight.

(*b*) A selector cock is fitted behind the loop aerial controls to enable the desired pump to be used. No. 1 position selects port inner engine pump.

(ii) *Pressure system:*

(*a*) A compressor on the port inner engine charges the wheel brake system reservoir. On later aircraft, the landing lamps are raised and lowered by this system.

(*b*) A compressor on the starboard inner engine operates the auto-pilot and Mark XIV bombsight.

(*c*) Two storage bottles, with valves for ground recharging, are fitted underneath the second pilot's seat.

PART I—DESCRIPTIVE

AIRCRAFT CONTROLS

8. **Rudder bars.**—The pedals are adjustable by means of a star wheel (33), the top of which should be pushed forward to increase leg reach.

9. **Trimming tabs**

Elevator:	Two crank handles (78) in roof, with indicator	Operate in natural sense
Rudder:	Crank handle (80) in roof, with indicator	Operates in natural sense
Aileron:	Fixed tab on each aileron	

10. **Automatic controls.**—The Mk. IV controls are grouped together on a panel (63) beside the port pilot's seat; *see* A.P. 2095, for operation of these controls.

11. **Wheel brakes.**—The lever (55) beside the throttle levers operates the wheel brakes; the degree of braking is controlled by varying the movement of the lever. Differential braking is controlled by the rudder bars.

12. **Undercarriage controls and indicator**

(i) The lever (60) in the left-hand slot of the control box has two positions only, RAISE and LOWER, with a safety catch (61) to retain the lever in the lower position. The switch (22) on the instrument panel has three positions, UP, OFF and DOWN, and is left at OFF except when undercarriage operation is required. The complete operation requires about one minute.

(ii) The indicator (17) consists of a set of lamps on the intrument panel comprising one pair, red and green, for each of the main wheel and tail wheel units. The switch (35) for the indicator is interconnected with the ignition switches. A dimming switch and reserve set of lamps are provided. The lights indicate as follows:

All units locked up .. .. 3 red lights

All units down and safe .. 3 green lights only

PART I—DESCRIPTIVE

(iii) Later aircraft have a standard type indicator indicating as follows:

All units locked up	No lights
All units unlocked	3 red lights
All units locked down and safe	3 green lights

NOTES.—

(i) When operating the undercarriage the lever must be set to RAISE or LOWER before operating the switch.

(ii) When setting to RAISE the lever must be pushed firmly in and when setting to LOWER it must be pulled firmly out.

(iii) The switch must be pushed hard over to ensure contact.

(iv) If it is necessary to switch off the motors during operation, 30 seconds must be allowed to elapse to ensure that the motors have stopped running before switching on again in the same, or the reverse, direction. Failure to observe this precaution may result in the clutch failing to engage and in damage to the motor.

(v) If it is necessary to reverse the direction of operation during raising, the wheels must be allowed to come up about half-way before the motor is switched off; otherwise, lowering electrically will not be possible.

(vi) If the red and green lights for any one wheel show together, with the wheel either up or down, a sticking limit switch or light switch is indicated. If this occurs:

(*a*) Before flight.—The aircraft should not be taken off until the trouble has been rectified.

(*b*) Before landing.—Go through the normal operation in order to lower the other wheels. Then lower the remaining wheel manually (this will not always be necessary, as on later aircraft the wheel may come down—although the light switch has stuck); but, as the green light was on before lowering, it is useless as an indicator and the wheel must be checked after lowering (electrically or manually) by the warning horn and also:

In the case of a main wheel, visually. On later aircraft check that the revolution counter fitted to undercarriage motors read the same as the reading before take off.

PART I—DESCRIPTIVE

In the case of the tail wheels at the jack as follows:
Nut at front of jack .. Tail wheels down
Nut at rear of jack .. .. Tail wheels up

13. **Undercarriage warning horn.**—This sounds if the flaps are more than one-third out when any wheel is not down.

14. **Flaps control and indicator.**—The flap operating switch (83) has three positions, IN, OFF, and OUT, and is mounted on a control panel above the instrument panel. Above it are the switch (84) for the flap indicator (87) and a tell-tale lamp which shows when the indicator is on. A red warning lamp (86) lights when the flaps are more than one-third out.

ENGINE CONTROLS

15. **Throttle and mixture controls**

(i) Throttle controls:

Mk. I

(*a*) On Mk. I aircraft four throttle levers (56) are mounted in the control box. The controls are of the "Exactor" type and must be primed before starting and in flight as follows: Push the levers fully forward, past the take-off position, hold them there for about 10 seconds, and then return them very slowly to the closed position.

Mk. III & IV

(*b*) On later aircraft the controls are mechanical and no priming is required; adjustable stops at the "closed" end of the slots being provided.

(ii) Mixture controls:

Mk. I

(*a*) On Mk. I aircraft two mixture levers (59), one controlling the port engines and one the starboard engines, are below the throttle levers. The controls are of the "Exactor" type, and are primed before starting and in flight by holding them fully down, past the NORMAL position, for about 10 seconds. Priming of the mixture controls in flight every 20 minutes or so is important to ensure correct functioning of the mixture control. Each mixture control is interlocked with the throttles on the same side so that the mixture lever can only be at weak (ECONOMICAL) when both throttles are in the cruising range.

PART I—DESCRIPTIVE

Mk. III & IV

(*b*) On later aircraft these are also mechanical, four levers being fitted, one for each engine. No priming is necessary.

(*c*) With Hercules XVI engines the mixture lever is not used and is inoperative; mixture regulation is automatic, and an economical mixture strength is obtained at or below plus 2 lb./sq.in. boost. For weak mixture cruising the throttle lever should not be opened beyond economical cruising position or beyond the white line if one is painted on the quadrant.

16. **Propeller speed controls**

Mk. I

(i) On Mk. I aircraft the levers (7) are in the lower part of the control box and are moved up to INCREASE and down to DECREASE, the R.P.M. The propeller speed controls are of the "Exactor" type, and should be primed before starting and in flight by holding them to the fully down position for about 10 seconds.

Mk. III & IV

(ii) On later aircraft the controls are mechanical, no priming is necessary and stops are fitted as in the case of the throttle levers.

17. **Propeller feathering controls** (40) are at the bottom of the instrument panel on the port side. For operation *see* Part IV.

18. **Supercharger controls.**—Four handwheels (93) are mounted on the front spar frame, two on each side, for controlling the two-speed superchargers. In the S gear position the pointer at the bottom of each wheel is turned inboard; the controls should be operated smartly.

19. **Carburettor air-intake heat controls.**—Four handwheels (90) are mounted above the supercharger controls, two on each side; they may be set to either of two positions, HOT or COLD.

20. **Cowling gills.**—The switches for these are on the engineer's panel. An indicator and warning light is fitted above each switch to show the position of the gills. Cylinder temperature gauges are on the engineer's panel.

21. **Slow-running cut-out controls.**—There are two levers (88) in a central position in the roof of the cockpit, the left-hand lever controlling the inboard engines and the other the outboard engines.

PART I—DESCRIPTIVE

22. **Priming pumps**

(i) The carburettor priming pumps are fitted in the centre section.

(ii) The induction priming system comprises Ki-gass priming pumps fitted in the centre section.

23. **Booster-coil switches.** — These switches (32) are fitted at the bottom of the pilot's instrument panel on the starboard side.

24. **Electric starter switches.**—The four engine-starter push-button switches are fitted below the ignition switches in a recess on top of the control box in the centre of the cockpit; they are protected from inadvertent operation by means of a spring-loaded cover plate (36). Provision is also made for hand turning the engines.

OTHER CONTROLS

25. **Bomb doors.**—Before bombs can be released, the trailing aerial must be wound in, fairlead retracted, and the doors under the bomb cells in the fuselage and/or main plane must be fully open. These doors are operated electrically and are controlled by two switches (27) on the right-hand side of the instrument panel, the left-hand switch controlling the fuselage doors and the other the main plane doors. Indicator lamps above the switches are illuminated when the doors are fully open, a duplicate set of lamps being fitted at the bomb aimer's station. In the event of electrical failure, the doors may be operated manually from the centre section.

26. **Bomb releasing.**—The selection, fusing and releasing of the bombs are carried out by the bomb aimer. The bomb release circuit is only complete when the appropriate bomb cell doors are fully open and, on early aircraft, when the main switch of the type "F" jettison switch on the starboard side of the instrument panel is in the ON position.

If necessary, any bombs selected by the bomb aimer can be released by the pilot by a duplicate firing switch at (52) on the port control handwheel, with a socket for connection (53) on the column.

PART I—DESCRIPTIVE

27. **De-icing.**—A de-icing solution may be sprayed on the pilot's windscreen and bomb aimer's window by means of a handpump mounted on the floor beside the captain; a cock at the bomb aimer's station enables this supply to be cut off if not required.

28. **Cabin heating controls.**—Handwheels (94) at the engineer's station control the valves of the port and starboard heating systems. On Mk. I (steam heating system only) switches by the valves control circulating pumps and must not be turned on before the valves have been set to HOT.

29. **Cockpit lighting.**—Small floodlights (14) with dimmer switches beside them illuminate the instruments.

30. **Radio controls.**—A remote control for the T.R.9.F. with a frequency change-over switch (48) is on the port side.

31. **I.F.F.**—Two pushbuttons shielded by a hinged flap and a master switch (30) are on the starboard side of the instrument panel.

32. **Beam approach.**—A panel (73) carries control and mixer switches. The indicator (15) is above the flying instrument panel.

33. **D.R. Compass.**—The switches (21) are at the centre of the instrument panel with the repeater (26) on the starboard side of the panel.

34. **Navigation and recognition lights.**—Switches (68) and (72) control these, the signalling switchboxes (76) being mounted above them. Indicator lamps (62) show when the respective lights are on. (On later aircraft indicator lamps are on the port side of wireless operator's station.)

35. **Landing lamps**

(i) On earlier Mk. I aircraft two landing lamps are mounted in the leading edge of the port wing, outboard of the outer nacelle; the "two-way and off" control switch (37) being mounted on the port side of the throttle lever quadrant. The lamps may be dipped by "Exactor" control by means of a lever to the right of the mixture levers.

PART I—DESCRIPTIVE

(ii) On later aircraft, one lamp is fitted in the same position in a box which can be lowered pneumatically by a similar lever (58) to that which previously dipped the lamps. The lamp box extends almost immediately after moving the lever and retracts in 4 to 5 seconds; after this time, the lever should be returned to the neutral position to cut off the air supply. A lamp dipping lever (57) is mounted beside the box control lever, and, to avoid overstraining, should not be operated whilst the lamps are retracted.

36. **Glider towing gear.**—On later aircraft a glider towing hook and remote control release are fitted; the release control is a long lever mounted on the starboard side of the throttle control box.

PART II

HANDLING

NOTE.—The speeds given in the following paragraphs as well as in Parts III and IV apply, except when otherwise stated, for either pressure-head position as well as when the A.S.I. is connected to the static vent.

37. **Fuel System Management**

The suggested sequence of use of tanks and management of the fuel system is set out in the charts and notes on the following pages. These should be read in conjunction with the diagram given below.

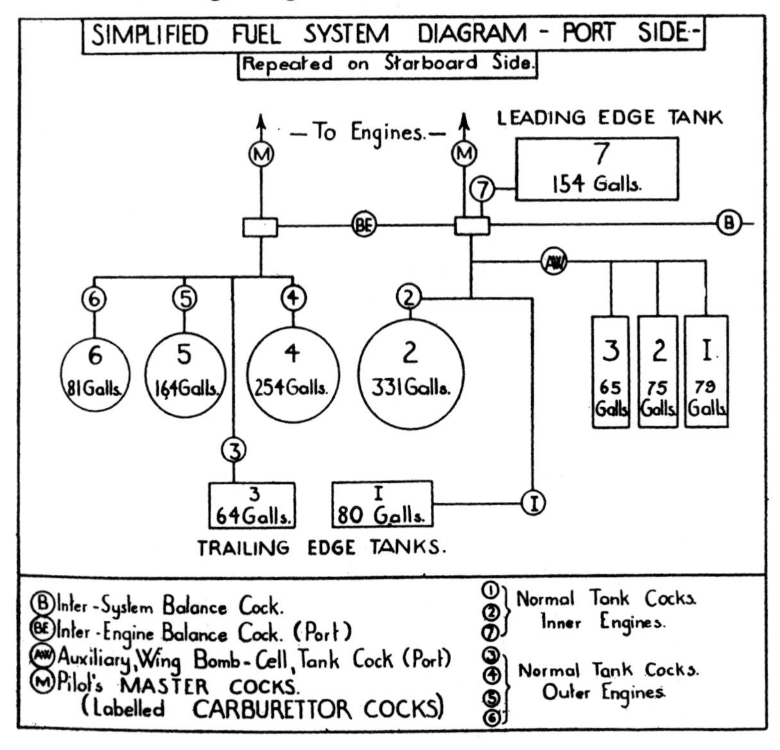

PART II—HANDLING

(i) *Normal system operation chart:*

The following table gives the recommended procedure when starting with the normal full fuel load of 2,254 gallons and assuming that 45 gallons are used for each engine for starting, take-off, and initial climb to about 3,000 ft.

	Inner Engines		Outer Engines	
	Tank and Cock	Fuel remaining	Tank and Cock	Fuel remaining
Starting T.O. and steep climbing	2 ON	331 galls.	4 ON	254 galls.
	2 OFF	286 "	4 OFF	209 "
Gentle climb and Level Flight	7 ON	154 "	3 ON	63 "
		101 "	3 OFF	10 "
		101 "	6 ON	81 "
		31 "	6 OFF	10 "
		31 "	5 ON	164 "
	7 OFF	10 "		143 "
	1 ON	80 "		143 "
	1 OFF	10 "		73 "
	2 ON	286 "		73 "
		223 "	5 OFF	10 "
		223 "	4 ON	209 "

NOTE:

The above sequence in use of tanks is recommended for the following reasons:

(*a*) Nos. 1 and 3 tanks should not be used for starting, take-off or steep climbing (as when taking evasive action, etc.) because in a tail down attitude they are below the level of the fuel pumps and there is a risk of fuel starvation. They should, however, be used early in flight to prevent the CG moving too far aft.

PART II—HANDLING

(*b*) No. 7 tanks should be used early in flight as they are not self sealing.

(*c*) If it is essential that maximum manœuverability be attained as early in flight as possible, Nos. 6 tanks may be used before Nos. 3, but Nos. 3 should, in any case, be used before any bomb load is dropped for C.G. reasons in (i)(*a*).

(ii) *Operation Chart using auxiliary wing bomb cell tanks:*

Total capacity of auxiliary wing bomb cell tanks is 438 gallons so total fuel, together with all normal tanks, is increased to 2,692, the following table gives the recommended procedure assuming that 45 gallons are again used for starting, take-off and initial climb to 3,000 feet.

	Inner Engines		Outer Engines	
	Tank and Cock	Fuel remaining	Tank and Cock	Fuel remaining
Starting T.O. and steep climbing	2 ON 2 OFF	331 galls. 286 ,,	4 ON 4 OFF	254 galls. 209 ,,
Gentle climb and Level Flight	Inter Engine Balance Cocks (BE) on 7 On 154 galls. 7 OFF— 10 ,,			
	Aux—ON	219 galls. 166 ,,	3 ON 3 OFF	63 galls. 10 ,,
		166 ,,	6 ON 6 OFF	81 ,, 10 ,,
	Aux—OFF	95 ,, 10 ,,	5 ON	164 ,, 79 ,,
	1 ON 1 OFF	80 ,, 11 ,,	5 OFF	79 ,, 10 ,,
	2 ON	286 ,,	4 ON	109 ,,
	EQUALISE—(BE)—ON then when equal—OFF	say 245 galls.		245 ,,

PART II—HANDLING

NOTES:

(*a*) The method set out above enables numbers 3, 5 and 6 tank gauges to be used to estimate the fuel in the auxiliary tanks; allowance should be made for any excessive consumption.

(*b*) When using auxiliary tanks in the above manner, all balance cocks should be OFF.

(*c*) If there is any doubt as to the fuel remaining in the auxiliary tanks, shut them off on one side and drain right down on the other side, i.e. until the fuel pressure warning lamp flashes or engine cuts. The time for this operation gives the approximate duration of fuel remaining in auxiliary tanks on the other side.

(iii) *General Notes:*

(*a*) If it is desired to use all possible fuel from each tank, wait until warning light comes on but then change over immediately. If contents gauges indicate the probability of two tanks on one side running dry simultaneously, change over on one while say 10 gallons still remain; this tank can be emptied later, after changing over on the other.

(*b*) No tank should be used for more than one engine at a time over enemy territory. Whenever one tank on each side is being used for both engines on that side, avoid allowing both to run out simultaneously as in (*a*) above.

(*c*) When there is a possibility of having to ditch the aircraft, it is advisable to leave 50 gallons in each No. 5 tank to use while ditching so that the contents of Nos. 2 and 4 tanks can be completely jettisoned. If all tanks excepting Nos. 2 and 4 are empty, in the event of ditching use the contents of one tank with the inter engine balance cock on, and jettison the contents of the other.

(*d*) Towards the end of the flight, the contents of Nos. 2 and 4 tanks should be equalised by opening inter engine balance cocks.

PART II—HANDLING

(*e*) For landing, should fuel be very short, run on No. 4 tanks—inter engine balance cocks on—until 10 gallons remain; then open No. 2 tank cocks and close balance cocks. When fuel pressure warning lights indicate that No. 4 tanks are running dry, close No. 4 tank cocks and open inter engine balance cocks. Allow 5 minutes between changeover on port and starboard sides to avoid two engines cutting simultaneously. The sequence of using Nos. 4 and 2 tanks may be reversed if defective gauges make this desirable.

(*f*) If one or more gauges become unserviceable the consumption from the tanks concerned can be checked by comparison with another tank feeding an engine running at the same boost and r.p.m.

(*g*) Numbers 2 and 4 tanks should always be turned on when the fuel system is in imminent danger of being damaged, i.e. in fighter belts, target area, etc.

(*h*) Tank cock positions are not marked on all aircraft but are as follows:

Levers up .. .. OFF
„ down .. .. ON

NOTE.—On Stirling III and IV aircraft the No. 7 tank controls operate in the opposite direction.

Levers up .. .. ON
„ down .. .. OFF

38. **Preliminaries**

Pilot's Checks

(i) Check:

Pilot's master cocks ..	OFF
Undercarriage master switch	OFF
Undercarriage lever ..	LOWER (switch on indicator and check green lights)
Bomb-doors	CLOSED
I.F.F.	OFF

PART II—HANDLING

D.R. Compass .. .. ON and SETTING
Trim tabs (rudder and elevator) for full movement and reset .. .. .. NEUTRAL
Pilot's escape hatch .. .. Fastened

(ii) Prime "Exactor" controls (Stirling I) as follows:

(*a*) Throttles .. Fully open, wait 10 secs., fully closed

Mk. I only.

(*b*) Mixture .. Fully down, wait 10 secs., NORMAL

(*c*) Propellers .. Fully down, wait 10 secs., fully up

Flight Engineer's Checks.—See A.P. 2764 Flight Engineer's Notes General and:

(i) Check that the jaws of the undercarriage " UP " locks are open.

(ii) Check tail chassis clutch pins are fully forward in the engaged position—emergency handle correctlv stowed. See that the electrical conduit to the motor is securely screwed home.

(iii) Check that main undercarriage motors are set for electrical operation—emergency handle stowed correctly —Red star wheel wound fully anti-clockwise (looking forward).

(iv) Check that flap operating motor is correctly set for electrical operation—emergency handle stowed correctly.

(v) Set air intakes to COLD, blowers to LOW (M).

(vi) Put 2 and 4 tanks " ON ", all other tanks " OFF ", port and starboard.

NOTE.—Nos. 2 and 4 tanks (port and starboard) should always be used for starting, running up, take-off and landing.
Balance cocks SHUT.

(vii) Check that all fuses are in position on the charging and distribution panel. (N.B.—Pay particular attention to the undercarriage fuse, which should be *OUT*.)

(viii) Check tension of undercarriage operating cables.

(ix) Vacuum pump to No. 2 position.

PART II—HANDLING

(x) Check inter-comm. at panel and bunk positions.

(xi) Check parachute stowage and security of all loose articles of equipment.

(iii) Flight engineer should also check with pilot that all actions in Pilot's List under (i) and (ii) have been carried out as well as all usual checks.

39. **Starting engines and warming up**

(i) Normally engines should be started in turn commencing with the port outer, unless starting from aircraft batteries, when start inner engines first. Before starting each engine, the following should be checked by the pilot and rechecked by the engineer visually and verbally:

(*a*) Check engines have been turned two revolutions by hand within the previous six hours.

(*b*) 2 and 4 tanks ON. All balance cocks OFF.

(*c*) COLD air, " M " Blower, gills OPEN.

(*d*) Carburettor cocks ON.

(*e*) Pilot should turn over engines to be started at least two revolutions with the starter (unless starting on aircraft batteries when engines should be turned over by hand) with throttle 1 in. open.

(ii) Start engines in turn as follows:

(*a*) Carburettors should not be primed, unless engines have not been run for one week or more, when engineer should prime with four single strokes of the wobble pump. If carburettors have been drained, nine single strokes should be given. *See* Part V.

(*b*) Engineer should prime induction system by working ki-gass pump, until the suction and delivery pipes are full; this may be judged by a sudden increase in resistance. High volatility fuel (stores ref. 34A/111) should be used, if an external priming connection is fitted, for priming at air temperatures below freezing.

(*c*) Pilot should switch ON ignition and booster coils, press starter button and call " contact " to engineer, who will work the priming pump as firmly as possible while the engine is being turned. It should start after the following number of strokes:

PART II—HANDLING

Air temperature °C	+30	+20	+10	0	−10	−20
Normal fuel ..	3	4	7	10		
High volatility fuel				3	8	18

If K.40 (40 c.c.) pumps are fitted, divide above number of strokes by four, giving an incomplete stroke where necessary.

Turning periods should normally be restricted to 10 seconds and must never exceed 20 seconds, with a 30-second wait between each.

(*d*) It will probably be necessary to continue priming after the engine has fired and until it has picked up on the carburettor.

(*e*) When the engine is running satisfactorily, switch OFF the booster coil (or starting magneto) and instruct the flight engineer to turn OFF the priming cock and screw down the priming pump.

(*f*) Run the engine as slowly as possible for half a minute, then warm up at about 1,000 r.p.m. to allow oil to circulate through the engine.

(*g*) After all engines have been started, flight engineer should check external supply disconnected and GROUND/FLIGHT switch set to FLIGHT.

.o. Testing engines and installations

(i) If the aircraft is facing out of wind, signal chocks away, turn into wind and apply brakes.

(ii) *While warming up :*

(*a*) Keep gills open and check with engineer the usual temperatures, etc.

(*b*) Test operation of controls, switch on flap indicator and test flaps. Flight engineer should check operation of flap motor while pilot is testing.

(*c*) Check brake pressure, at least—100 lb./sq.in.

iii) *After warming up :*

The following comprehensive checks should be carried out after repair, inspection (other than daily), or otherwise at the pilot's discretion. Normally they may be reduced in accordance with local instructions.

PART II—HANDLING

(*a*) At 1,500 r.p.m. with propeller lever fully up, order the engineer to put in " S " gear. Note a momentary drop in oil pressure (engineer) and a slight flicker in boost (pilot). Change back to " M " gear.

NOTE.—This test need only be carried out once per day.

(*b*) Open engine up to 2,400 r.p.m. and set propeller lever fully down to check speed variation of governor through full range; drop should be 600/800 r.p.m. A drop of say 200 r.p.m. is, however, sufficient to indicate that the C.S.U. is functioning. Return lever fully up. Ensure that the r.p.m. returns to 2,400.

(*c*) When running up inners, check instruments, and the engineer should check the charging rate of the generators.

(*d*) Check functioning of mixture control (except Hercules XVI); at −1 to −2 lb./sq.in. R.p.m. should drop slightly.

(*e*) If necessary to check static boost and r.p.m., open throttle slowly to the take-off position, noting the boost reading of $+8\frac{1}{4}$ at 2,800 to 2,850 r.p.m. ($+6\frac{3}{4}$ and 2,800 r.p.m. with Hercules XI).

(*f*) With propeller levers set fully up, open up to rated boost. If the propeller is constant speeding at this boost, the throttle should be closed until a slight drop in r.p.m. is noted. Test magnetos: maximum drop 50 r.p.m.

(*g*) While throttling back, check boost at climbing and cruising stops.

(*h*) Close throttle and check slow running: 400/600 r.p.m.

(*i*) Open throttle to obtain 1,000 r.p.m. approximately.

NOTE.—(*a*) Ground running should be reduced to a minimum; by co-operation with ground crew, one run-up can sometimes take the place of two.
(*b*) Throttles must not be opened quickly and engines should not be run at high boost and r.p.m. for longer than is absolutely necessary.

41. **Taxying out**

(i) Do not run engines at less than 1,000 r.p.m., using mainly outer engines for manœuvring on the ground.

PART II—HANDLING

(ii) Signal chocks away, set inners at 1,000 r.p.m. and taxy to the take-off position. Ensure the brake pressure is sufficient to reach the take-off point—min. 120 lb./sq.in.

(iii) When turning, the aircraft must on no account be allowed to pivot about one wheel, or considerable damage to tyres may result.

(iv) After clearing engines prior to take-off, check that the throttles are synchronized for slow running. If not, prime " exactor " control as necessary.

Mk. I.

WARNING.—When fitted, the under-turret guns tend to creep into the fully-down position, when they may foul the ground. The air gunner should, therefore, ensure that the guns are fully elevated before (and if no stop or catch is fitted remain so during) taxying, take-off and landing.

42. **Check lists before take-off**

(i) *Pilot checks*

T—Trimming tabs:	
Rudder tab:	Neutral
Elevator tab:	
(*a*) C.G. forward	Neutral
(*b*) C.G. normal (111 to 112 ins. aft) ..	4 divisions forward
(*c*) For each 5 ins. aft of (*b*)	4 divisions further forward
M—Mixture	NORMAL
P—Propeller	Speed controls fully up
F—Fuel	Check master cocks, also tank cock settings and contents with engineer
F—Flaps	One-third out
Superchargers	LOW. Check with engineer
Gills	One-third open. Check with engineer
D.R. compass	NORMAL

NOTE.—When setting flaps one-third out, set switch to IN when one-third warning light comes on, and to OFF immediately it goes out.

PART II—HANDLING

(ii) *Flight engineer checks*

(*a*) Superchargers LOW.

(*b*) Gills one-third open and check by indicator and visually that all are at same angle to engine cowlings.

(*c*) Check and inform pilot, temperatures and pressures normal, undercarriage fuses *IN*.

NOTE.—Desired temperatures for take-off:

Oil ..	.. 25° C.–70° C.
Cylinder ..	.. 160° C.–220° C.

43. Take-off

(i) Turn into the take-off direction, and run forward a few yards to straighten the tail wheels.

(ii) With brakes on, open each throttle until the engines are running at about 2,000 r.p.m. Then release the brakes and open the throttles slowly (holding all four levers with one hand) the starboard throttles leading, to counteract the slight tendency to swing to starboard. Ease the control column forward enough to lift the tail as speed is gained, then keep straight with the rudder.

(iii) If taking off at full load and with the C.G. 120 ins. or more aft, the elevator tabs should be wound fully forward as the undercarriage is rising. When the undercarriage is up, the aircraft will still be markedly tail heavy, but when the flaps are raised, trim is regained, the tab setting for the climb being 8 divisions forward.

(iv) (*a*) The aircraft should be eased off the ground at not less than the following speeds:

	Thousand lb.		
Weight	56–60	60–65	65–70
Take-off speed m.p.h. I.A.S.* ..	100	105	110

(*b*) The safety speed is 135 m.p.h. I.A.S.*

* On early aircraft which have the pressure head on top of the wireless mast add 5 m.p.h.

44. After take-off

(i) (*a*) Brakes on gently, undercarriage up, brakes off.

(*b*) Second pilot to check operation as wheels come up and report anything abnormal to engineer at once.

(*c*) When wheels are up, set switch OFF.

PART II—HANDLING

(ii) (*a*) Do not start to raise the flaps until the undercarriage is completely retracted. The flaps come up very slowly and there is little tendency to lose height.
(*b*) Before raising flaps the switch may be set OUT until ⅓ out lamp lights to check that warning buzzer operates, then set to IN to raise flaps.

45. **Climb**

When undercarriage and flaps are up, the aircraft may be climbed at 150 m.p.h. I.A.S. with the gills ⅓ open (if temperatures permit to not more than the "straight" position) or at about 185 m.p.h. I.A.S. with the gills closed.
See para. 57.

46. **General flying**

(i) *Stability.*—The aircraft is directionally and laterally stable. It is reasonably stable longitudinally except at full load with the C.G. in the most aft position.

(ii) *Change of trim*

Flaps down	Nose up
Undercarriage down	Nose down
Gills closed	Little change

(iii) *Controls.*—The controls are good, the ailerons being reasonably light for such a large aircraft. The elevator control has a slightly heavy and sluggish initial movement.

(iv) *Flying at low airspeeds.*—Flaps may be lowered to the ⅓ out position and the speed reduced to about 145 m.p.h. I.A.S.

(v) *Astro-navigation.*—The aircraft can be flown by experienced pilots under manual control with sufficient steadiness for astro-navigation, but it is recommended that the auto-control should be used while sextant observations are taken.

(vi) *Synchronising engines.*—The engines may be synchronised by r.p.m. indicators; a fine adjustment of the inners can usually be made by sound and then of the port and starboard pairs visually by looking through the propeller discs.

(vii) Desired engine temperatures for continuous flight:

Cylinder	200° C. to 220° C.
Oil	55° C. to 70° C.

PART II—HANDLING

47. Stalling

(i) Experience indicates that the I.A.S. at the stall varies considerably with different aircraft; only very approximate figures can therefore be quoted. At about 10 m.p.h. above the stall buffeting occurs, and at the stall the nose drops, followed by a slight tendency for the starboard wing to drop.

(ii) Approximate stalling speeds at 62,000 lb. (engines off) with underbody pressure heads are:

Flaps and undercarriage up ..	110 m.p.h. I.A.S.
,, ,, ,, down ..	85 m.p.h. I.A.S.

With pressure head on wireless mast, or with A.S.I. connected to static vents, about 10 to 15 m.p.h. higher.

48. Before landing

Flight Engineer's checks

(i) Check, as accurately as possible, the amount of fuel remaining and inform the pilot quantity and distribution.

(ii) Check that the all-up weight of the aircraft does not exceed 60,000 lb. for landing, by adding to the normal service weight (about 48,900 lb. Mk. I, 49,800 lb. Mk. III and IV without bombs or fuel) the weight of fuel remaining, i.e. gallons × 7·3 lb. At 49,800 lb. service weight, fuel should not exceed about 1,300 gallons. If A.U.W. is excessive, jettison fuel as necessary.

49. Approach and landing

(i) *Pilot's actions*

Reduce speed to about 145 m.p.h. I.A.S. and act as follows:

(*a*) Lower flaps to ⅓ out.

Mk. I. (*b*) Prime " Exactor " throttle and propeller controls (Mk. I only) and set latter to give 2,500 r.p.m.

(*c*) Check brake pressure with lever applied. If the system has been damaged, the gauge may give a false reading with the brakes off. Release brakes after check.

PART II—HANDLING

(ii) *Flight Engineer's actions*

Check superchargers ..	LOW
Gills	⅓ open (if necessary)
Carburettor air ..	COLD
Check fuel	*See* paras. 37 and 48.

(iii) *Check lists before landing*

(*a*) *Pilot*

U—Undercarriage ..	Operating lever LOWER Master switch DOWN When green lights appear, put master switch OFF
M—Mixture controls ..	NORMAL
P—Propeller	Speed controls 2,500 r.p.m. or fully up, for final approach
F—Flaps	Fully OUT for straight final approach
Under turret (if fitted)..	Guns elevated (*see* para. 41 WARNING)

(*b*) Flight engineer, as wheels come down, checks:
Main wheels, visually (and by revolution-counter on later aircraft); tail wheel, jack nut fully forward to tail wheel motor.
Gauge readings.

(*c*) If desired by pilot, second pilot should call out I.A.S. and height at intervals.

(iv) The recommended minimum approach speeds are:

	A.S.I. connected to:		
	Mast Pressure Head	Under Body Pressure Head	Static Vent
At 48,000 lb.:			
Engine asstd.	115	100	105
Glide ..	125	110	115
At 56,000 lb.:			
Engine asstd.	125	110	115
Glide	135	120	125

PART II—HANDLING

50. Mislanding

Opening the throttles with the flaps and undercarriage down makes the aircraft markedly tail heavy. After mislanding, open throttles slowly to maximum rich continuous boost, retrim the aircraft and then open the throttles fully. Raise the flaps to $\frac{1}{3}$ out (this produces no tendency to sink) and retrim as necessary, raise the undercarriage and proceed as for normal take-off. If propellers are set to 2,500 r.p.m., set fully up immediately should it be necessary to open up to take-off boost.

51. After landing

(i) Engineer opens gills fully before taxying.

(ii) If taxied for long distances, engines should be cleared by opening up to 2,000 r.p.m. with propeller controls fully up.

52. Stopping engines

(i) Turn into wind with tail wheels straight.

(ii) To reduce risk of fire in the induction system and to prevent subsequent hydraulicing, stop engines as follows:

(*a*) Run engine at 800 to 900 r.p.m. until cool.

(*b*) Open up slowly and evenly to 2 lb./sq.in. boost for 5 seconds.

(*c*) Close throttles slowly and evenly, taking about 5 seconds, until speed is reduced to 800 to 1,000 r.p.m. and run at this speed for 2 minutes, or longer if diluting.

(*d*) Pull carburettor cut-out and switch off when engine stops

(iii) *Oil dilution.—See* A.P. 2095 and note: the correct period is 4 minutes with engines running at 900 to 1,000 r.p.m.

53. When parking

See A.P. 2764 Flight Engineer's Notes General and note:

(i) Pilot and Engineer — All electrical switches OFF
GROUND/FLIGHT switch—GROUND
Remove u/c fuse

(ii) On Mark I aircraft, to relieve load on exactor controls, leave as follows:

Mk. I.

Throttles .. Fully OPEN (after engines have cooled off)
Mixture .. NORMAL
Propellers .. Fully down

AIR PUBLICATION 1660A, C & D—P.N.
Pilot's and Flight Engineer's Notes

PART III

OPERATING DATA

54. Engine data

(i) *Hercules XI* (*Operational*)

(*a*) Fuel .. .. 100 octane only

(*b*) Limitations:

		R.p.m.	Boost lb./sq.in.	Temp. ° C. Cylr.	Oil
MAX. TAKE-OFF TO 1,000 FT. ..	M	2,800	+6¾		
MAX. CLIMBING 1 HR. LIMIT ..	M, S	2,500	+3½	270 (290)	90
MAX. RICH CONTINUOUS ..	M, S	2,500	+3½	270 (290)	80
MAX. WEAK CONTINUOUS ..	M, S	2,500	+1	270 (290)	80
COMBAT .. .. 5 MINS. LIMIT..	M, S	2,800	+6¾	280 (300)	100

NOTE.—Use of the higher temperatures in brackets is permitted only when operational conditions make the observance of the normal limitations impracticable, but the life of the engine will be shortened.

OIL PRESSURE:
NORMAL .. .. 80 lb./sq.in.
EMERGENCY MINM. (5 MINS.) .. 70 lb./sq.in.

MINM. TEMP. FOR TAKE-OFF:
OIL .. .. .. 5° C. (recommended 15° C.)

MAX. TEMP. FOR TAKE-OFF:
CYLINDER .. .. 230° C.

MAX. TEMP. FOR STOPPING ENGINES:
CYLINDER .. .. 230° C.

PART III—OPERATING DATA

(ii) *Hercules VI and XVI* (*Operational*)

(*a*) Fuel .. .. 100 octane

(*b*) Limitations:

		R.p.m.	Boost lb./sq.in.	Temp. ° C. Cylr.	Oil
MAX. TAKE-OFF TO 1,000 FT. ..	M	2,800	$+8\frac{1}{4}$		
MAX. CLIMBING 1 HR. LIMIT ..	M S	2,400 2,500	+6	270 (290)	90
MAX. RICH CONTINUOUS ..	M S	2,400	+6	270 (290)	80
MAX. WEAK CONTINUOUS ..	M S	2,400	+2	270 (290)	80
COMBAT .. .. 5 MINS. LIMIT ..	M S	2,800	$+8\frac{1}{4}$	280 (300)	100

Temperatures in brackets are permitted when thermo-couples are fitted to No. 14 cylinder.

OIL PRESSURE:
NORMAL .. .. 80–90 lb./sq. in.
EMERGENCY MINM. (5 MINS.) .. 70 lb./sq. in.

MINM. TEMP. FOR TAKE-OFF:
OIL .. .. 5° C. (15° C. recommended)

MAX. TEMP. FOR TAKE-OFF:
CYLINDER .. .. 230° C.

MAX. TEMP. FOR STOPPING ENGINES:
CYLINDER .. .. 230° C.

(iii) *Hercules XI, VI and XVI* (*Training*)

The following limitations apply for use with 87 octane fuel in training service:

		R.p.m.	Boost lb./sq.in.	Temp. ° C. Cylr.	Oil
MAX. TAKE-OFF TO 1,000 FEET ..	M	2,800	+5		
MAX. CLIMBING 1 HR. LIMIT ..	M S	2,400	$+2\frac{1}{2}$	270	90
MAX. RICH CONTINUOUS ..	M S	2,400	$+2\frac{1}{2}$	270 (250)	80
MAX. WEAK CONTINUOUS ..	M S	2,200 (2,400)	Zero	270 (250)	80
MAX. ALL-OUT 5 MINS.	M S	2,800	+5	280	100

NOTE.—Figures in brackets apply to Hercules VI and XVI.

OIL PRESSURE:
NORMAL .. .. 80 lb./sq.in.
EMERGENCY MINM. (5 MINS.) .. 70 lb./sq.in.

MINM. TEMP. FOR TAKE-OFF:
OIL .. .. 5° C. (recommended 15° C.)

MAX. TEMP. FOR TAKE-OFF:
CYLINDER .. .. 230° C.

MAX. TEMP. FOR STOPPING ENGINES:
CYLINDER .. .. 230° C.

PART III—OPERATING DATA

55. **Position error corrections**

(i) On aircraft with the pressure head on top of the aerial mast, the corrections are as follows:

(*a*) At 65,000 lb.:

	Flaps Down		Flaps Up								
From ..	100	120	120	135	150	165	180	200	215	230	m.p.h.
To ..	120	140	135	150	165	180	200	215	230	245	I.A.S.
Add ..	–	–	–	–	–	–	–	2	4	6	m.p.h.
Subtract	4	2	8	6	4	2	0	–	–	–	

(*b*) At 50,000 lb.:

	Flaps Down		Flaps Up								
From ..	100	120	115	130	145	160	175	190	205	220	m.p.h.
To ..	120	140	130	145	160	175	190	205	220	240	I.A.S.
Add ..	–	–	–	–	–	–	2	4	6	8	m.p.h.
Subtract	2	0	6	4	2	0	–	–	–	–	

(ii) On aircraft with the pressure head beneath the fuselage, the corrections are as follows:

(*a*) At 65,000 lb., flaps up:

From	120	130	145	160	180	200	220	240	m.p.h.
To	130	145	160	180	200	220	240	270	I.A.S.
Add	4	2	0	–	–	–	–	–	m.p.h.
Subtract	–	–	–	2	4	6	8	10	

(*b*) At 50,000 lb., flaps up:

From	120	130	140	155	170	190	210	240	m.p.h.
To	130	140	155	170	190	210	240	270	I.A.S.
Add	2	0	–	–	–	–	–	–	m.p.h.
Subtract	–	–	2	4	6	8	10	12	

(iii) When connected to the static vents the I.A.S. position error is less than 2 m.p.h. throughout the speed range, and may, therefore, be neglected.

PART III—OPERATING DATA

56. Flying limitations

(i) This aircraft is designed for duty as a heavy bomber and intentional spinning and aerobatics are not permitted.

(ii) *Maximum speeds in m.p.h. I.A.S. :*

Diving: Pressure head under fuselage	.. ..	325
,, ,, on mast	.. .. ..	295
A.S.I. connected to static vents	.. ..	310

Undercarriage locked down	155	With pressure head on top of mast, add 5 m.p.h.
Flaps down	145	
Landing lamp lowered ..	155	

(iii) *Maximum weights :*

Take-off and straight flying only	70,000 lb.
Landing and all forms of flying	60,000 lb.

(iv) *Bomb clearance angles :*

Dive	30°
Climb	20°
Bank	8° (with S.B.C. 11°)

57. Maximum performance

(i) *Climbing:*

(*a*) Speed for maximum rate of climb in m.p.h. I.A.S.:

From S.L. to 10,000 ft.	150
Above 10,000 ft.	145

NOTE.—Experience indicates that under normal conditions an equally good rate of climb, with somewhat lower oil temperatures, can be maintained by climbing at 180/185 m.p.h. I.A.S. with gills closed.

(*b*) Change to S gear when the boost has fallen by 3 lb./sq.in.

(ii) *Combat:*

Use S gear if the boost in M gear is less than 3 lb. below combat boost limit.

PART III—OPERATING DATA

58. **Maximum range** (*see* curves, page 36.)

(i) *Climbing:*

(*a*) Climb at 150 m.p.h. I.A.S. using maximum boost and r.p.m. (+3½ lb./sq.in. and 2,500 r.p.m. with Hercules XI and +6 lb./sq.in. and 2,400 r.p.m. with Hercules VI)—*see* para. 57(i).

(*b*) The climb may be made in stages as follows: Level off, and then reduce to maximum weak mixture cruising conditions. The speed which can then be maintained is about 155 to 165 m.p.h. I.A.S. When resuming the climb, change to climbing conditions, if necessary flying level until the desired climbing speed is reached.

(*c*) Above full throttle height follow the boost back with the throttle levers and change to S gear only when boost has fallen by 3 lb./sq.in.

(*d*) Delay the climb above full throttle height as late as possible until the aircraft is lighter. The I.A.S. should be reduced progressively to maintain a reasonable rate of climb and should be about 145 m.p.h. I.A.S. at about 17,000 ft.

(ii) *Cruising:*

(*a*) The recommended speeds are:

(i) Fully loaded, outward journey: 165 m.p.h. I.A.S. (when fully loaded it may not be possible to maintain more than 155–160 m.p.h. I.A.S.).

Mk. I (ii) Lightly loaded, homeward journey: 160 m.p.h. I.A.S.

(*b*) Fly in M gear and weak mixture with throttles set to give just under +1 lb./sq.in. (+2 on Hercules VI) boost, and adjust r.p.m. to give the recommended air speed.

(*c*) Engage S gear if the recommended speed cannot be maintained in M gear at 2,500 r.p.m. (2,400 on Hercules VI), but only when the boost has fallen to −1 lb./sq.in. (zero on Hercules VI).

(*d*) With Hercules XI engines, r.p.m. in excess of 2,300 in S gear should be avoided; a lower I.A.S. being accepted.

PART III—OPERATING DATA

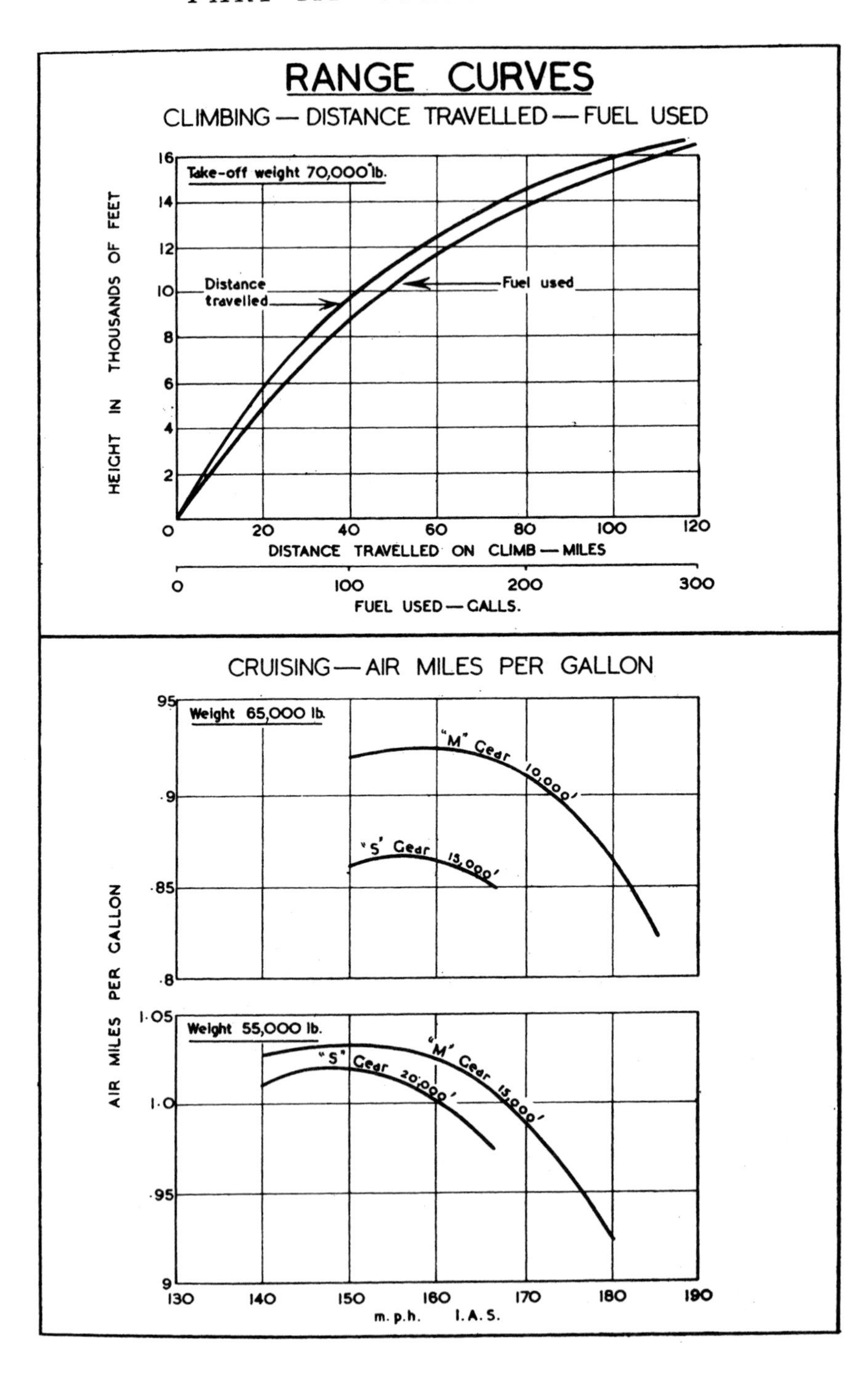

PART III—OPERATING DATA

59. **Use of warm and cold air**

See A.P. 2095 P.N. General and note that use of warm air may result in a slight reduction in range.

60. **Fuel capacities and consumptions**

(i) Normal fuel capacity excluding No. 7 tanks 1,946 galls.
,, ,, ,, including No. 7 tanks 2,254 galls.
Auxiliary (wing bomb cell) tanks .. .. 438 galls.
Total fuel (excluding No. 7 tanks) .. 2,384 galls.
,, ,, (including No. 7 tanks) .. 2,692 galls.

(ii) The approximate consumptions for the aircraft in rich mixture are:

R.p.m. and boost for:	Gallons/hour Herc. XI	Gallons/hour Herc. VI
Max. climbing	420	480
Max. rich continuous	420	480
Combat	580	640

(iii) The approximate consumptions for the aircraft in a weak mixture in gallons per hour are:

(*a*) With Hercules XI engines:

Boost	M ratio at 5,000 ft.				S ratio at 15,000 ft.			
	r.p.m.				r.p.m.			
lb./sq.in.	2,500	2,400	2,200	2,000	2,500	2,400	2,200	2,000
+1	(254)	(248)	(226)	(186)	(236)	(230)	—	—
0	(234)	228	214	192	(228)	222	208	—
−1	(222)	216	198	182	(218)	214	196	184
−2	(208)	202	186	164	(210)	202	184	174
−3	(192)	198	172	154	(200)	192	174	162
−4	(176)	170	158	142	(192)	180	162	150

For each 2,000 ft. above or below the altitudes quoted add or subtract 2 gallons/hour. The figures in brackets are estimated only.

PART III—OPERATING DATA

(*b*) With Hercules VI engines:

Boost lb./sq.in.	M gear at 5,000 ft. r.p.m.			S gear at 15,000 ft. r.p.m.		
	2,400	2,200	2,000	2,400	2,200	2,000
+2	235	220	204	236	226	216
0	212	196	184	212	204	196
−2	188	176	164	192	184	176
−4		160	148		128	160

Add or subtract for each 1,000 ft. above or below the above heights:

M gear .. .. .. 1 gallon per 2,000 ft.
S gear .. .. .. 1 gallon per 1,000 ft.

(iv) Air miles per gallon.—The curves on page 36 show the air miles per gallon for typical outward and homeward loads, at various I.A.S.

AIR PUBLICATION 1660A, C & D—P.N.
Pilot's and Flight Engineer's Notes

PART IV

EMERGENCIES

61. **Engine failure during take-off**

(i) Failure of one engine.—At loads up to 58,000 lb. the aircraft will climb at 145 m.p.h. I.A.S. and climbing boost. Do not climb at less than 145 m.p.h. I.A.S. With pressure head on mast add 5 m.p.h. Leave the flaps at the take-off setting until a safe height has been reached and until the undercarriage has been raised.

NOTE.—Pilots with considerable experience of the type *may* find it possible to climb away at loads up to 62,000 lb. Performance is affected by the position of the undercarriage at the time of engine failure, and depends upon which engine fails (e.g. failure of the starboard outer engine has the worst effect).

(ii) Failure of two engines. It will be necessary to close the throttles and land.

62. **Engine failure in flight**

(i) With any one engine out of action, the aircraft can maintain height at any load.

(ii) With two engines out of action, difficulty will be experienced in maintaining height at any load over about 50,000 lb., and therefore jettisoning may be necessary.

(iii) The fuel supply to a failed engine should be shut off by the pilot at the master cock after feathering.

(iv) A speed of 145 m.p.h. I.A.S. (or more if foot load with full trim is excessive) should be maintained.

63. **Feathering**

(i) Press feathering button and let go.

(ii) Close throttle immediately.

(iii) If engine is to be out of action for any length of time, after it has stopped switch off and turn off master fuel cock.

PART IV—EMERGENCIES

64. Unfeathering

(i) Set propeller speed control fully down and throttle closed or slightly open.

(ii) Turn on ignition and master fuel cock (if off).

(iii) Press feathering button and hold until r.p.m. reach 1,000 to 1,300. If the propeller does not return to normal constant-speed operation, open throttle slightly.

65. Undercarriage emergency operation

(i) Before operation, the pilot must set the master switch to OFF.

(ii) The undercarriage can then be operated manually by a member of the crew as follows.

(iii) *Lowering main wheels:*

(Later aircraft with motors in fuselage)

(*a*) Check with pilot, master switch OFF.

(*b*) Remove the locking pin and rotate the red lock-release star wheel clockwise (looking forward) until the limit stop is reached; this ensures that the undercarriage locks are released and that, in the event of failure at an intermediate stage, it will not be necessary to operate this control.

(*c*) Hinge down the transparent cover.

(*d*) Pull out the *green* clutch body and twist so that the notch against the flap on the body is in line with one of the notches on the internal barrel; this disengages the motor from the gearbox and ensures that when fitted, the handle engages correctly.

(*e*) Fit the handle so that its lug engages against the flat on the clutch body and turn the handle until the *green* light shows and the revolution counter reaches 0 (the switch for the revolution counter lamp is on the fuselage side forward of the gearbox and motor).

(*f*) Turn as follows:

Port wheel .. .. Anti-clockwise
Starboard wheel .. .. .. Clockwise

PART IV—EMERGENCIES

(iv) *Lowering main wheels:*

(Early aircraft only)

NOTE.—The figures in brackets in the following paragraphs correspond with the figures stamped on the undercarriage control levers in the aircraft.

On aircraft with undercarriage manual control boxes (consisting of a cylindrical casing) fitted on either side of the fuselage aft of the top main rear spar truss.

(*a*) Remove locking pin from red star wheel (by lower rear spar on starboard side) and turn clockwise to full extent.

(*b*) Remove locking pins from the blue handle (2) and the green lever (3) on the undercarriage control box.

(*c*) Exert pressure on green lever (3) while rotating handle (2) back and forth, until green lever (3) is fully over to the " hand operation " position.

(*d*) If the green lever (3) does not come fully over to the " hand operation " position, maintain pressure, remove locking bolt from handle (4) and oscillate slightly; if necessary, continue to rock (2) until the lever (3) is fully over to the hand operation position.

(*e*) Turn handle (4) in the direction indicated by the arrow until the green light (5) appears; give additional turns as stated on the undercarriage instruction plate on the side of the fuselage near the emergency gear.

Some early aircraft have no undercarriage manual control boxes fitted, when proceed as follows:

(*f*) Remove locking pin from red star wheel and turn clockwise to full extent.

(*g*) Pull (1) inboard and lock by rotating.

(*h*) Remove locking pin from (2) and (3).

(*i*) Pull (3) inboard while rotating (2) back and forth from stop to stop.

(*j*) If (3) cannot be fully withdrawn, maintain pressure and oscillate handle (4); if necessary, continue to operate (2) from stop to stop until the disengaging pinion operated by (3) is fully withdrawn.

PART IV—EMERGENCIES

(*k*) When (3) is fully withdrawn, turn (4) in the direction indicated by the arrow until the green light (5) appears. Give additional turns as in (*e*) above.

(v) *Raising main wheels* (later aircraft)

Proceed as for lowering, but turn as follows:

Port wheel .. .. Clockwise
Starboard wheel .. .. .. Anti-clockwise

(vi) *Raising main wheels* (early aircraft)

This cannot be done manually.

(vii) *Lowering tail wheel* (all aircraft)

(*a*) Check with pilot, master switch OFF.

(*b*) Push rearwards the clutch pins projecting from the gearbox casing and twist clockwise to lock; this disengages the motor.

(*c*) Fit the handle and turn it anti-clockwise until the *green* light shows; continue turning with the number of turns specified on the instruction label.

NOTES.—(*a*) The approximate times for raising and lowering the units are as follows:

Later aircraft
Main wheels:
Lowering, 5½ mins.; raising, 14 mins.
Tail wheels:
Lowering, 4 mins.

Early aircraft
Main wheels:
Lowering, 8½ mins.
Tail wheels:
Lowering, 4 mins.

(*b*) Once manual operation has been commenced, do not revert to electrical operation without first resetting (*see* ix).

(viii) *Crew drill to be carried out during manual lowering of undercarriage:*

The following drill is recommended as the time taken for manual operation of the undercarriage can be considerably reduced thereby; this is a great advantage when returning from a long trip (little fuel left, etc.).

PART IV—EMERGENCIES

(*a*) The engineer should set the gear for manual operation on both main wheels and should start winding the wheels down. He should then instruct two other members of the crew, e.g. W.Op. and navigator, to continue winding the main wheels keeping the number of turns decreasing on the rev. counter gauges approximately the same. They should be instructed that should anything abnormal happen, both are to cease winding and await the engineer who, in the meantime, will go aft, set the tail wheels for manual operation and wind fully down.

(*b*) On return to the rear spar position, the engineer will inspect the rev. counters, and check or complete the locking of both main wheels himself.

(ix) *Resetting main wheels*

Later aircraft

(*a*) Ensure that the lock-release star wheel is fully wound anti-clockwise.

(*b*) Pull the operating handle straight out and stow.

(*c*) Twist the clutch body (either way) until it snaps back into position with the dogs fully engaged. This re-engages the motor ready for electrical operation.

(*d*) Hinge up the transparent cover over the clutch body.

Early aircraft

(*e*) The resetting operation cannot be carried out by the air crew in flight.

(x) *Tail wheel resetting (all aircraft)*

(*a*) Twist the clutch pins anti-clockwise and allow the internal spring to return them to the front of the horizontal slots; if necessary, rotate the handle slightly to assist the gears to engage and thus permit the pins to return.

(*b*) Remove the handle and stow.

66. **Flaps emergency operation**

In the event of electrical failure, a member of the crew should operate the flaps manually as follows:

PART IV—EMERGENCIES

(i) Push the knurled barrel beside the gearbox outward ar twist anti-clockwise (facing starboard); this disengag motor.

(ii) Fit the operating handle and turn as indicated.

67. **Bomb door emergency operation**

(i) In the event of electrical failure, a member of the crew should pull out the knob below the emergency position indicator lamps on the port side of the centre section (this brings them into operation) and proceed as follows:

Fuselage bomb doors

(*a*) Hinge back domed cover plate in the floor beside the gearbox.
(*b*) Push down the handle and twist to lock (on early aircraft pull out toggle and turn to lock).
(*c*) Fit the handle and rotate as required until the indicator lamp lights.

Wing bomb doors

(*a*) Pull out the toggle (S) on the centre section side and twist to lock.
(*b*) Fit the handle and rotate as required until the indicator lamp is illuminated.

NOTE.—Once manual operation has been commenced, do not revert to electrical operation without re-setting.

(ii) *Resetting*

Unlock the handle and allow the spring action to return it to its former position; if it does not return, rotate the handle slightly until it does. Remove the handle and stow.

68. **Damage by enemy action**

Engineer's checks:

The engineer should know the run of all control cables and tank cock cables as well as the location of all fuses, other accessories and equipment which may require attention. The following checks should be made:

PART IV—EMERGENCIES

(i) Check that engines are running on Nos. 2 and 4 tanks, check fuel remaining in each tank and note.

(ii) Check engine temperatures and all pressures and instruct another member of the crew to keep a watch on them.

(iii) Go to the rear turret and check elevator, rudder and trimmer control cables throughout their length for damage. Report to pilot.

NOTE.—Always carry a spare piece of cable, locking wire, and pliers. It does not matter how rough the repair on control cables is, so long as the pilot can obtain a certain amount of response from the control.

(iv) Check all turret pipe lines and recuperators.

(v) Check all oxygen equipment.

(vi) Inspect aircraft structure for serious damage.

(vii) Check batteries.

(viii) As far as possible, check the undercarriage and flaps for damage.

(ix) Should any fires be located, use asbestos gloves and hand extinguishers. Keep oxygen mask and goggles on when dealing with fires. Any detachable equipment such as pyrotechnics, boardings, covers, etc., should be jettisoned.

Do not hesitate—act immediately.

(x) Should a fire be located in the bomb bays, request pilot to open doors, jettison bombs and dive aircraft.

(xi) Have mid-upper gunner inspect carefully (if he is not otherwise engaged) the upper side of both wings for damage, oil, etc. This enables an engineer, who should know the exact position of the various components in his wings, to estimate more accurately what damage may have occurred and so anticipate trouble which may not come to light until later in the flight.

(xii) Engineer should return to his fuel gauges and check. Should loss of fuel become apparent he should act in accordance with the fuel system management instructions. *See* Part II and Part V.

PART IV—EMERGENCIES

(xiii) A constant check on fuel, engine temperatures and pressures, etc. should be kept.

(xiv) Check aileron controls from the control column to the point where they disappear into the wing roots.

NOTES:

(*a*) A partially frayed cable is a danger—reinforce with spare piece of cable and bind together with locking wire.

(*b*) A broken rod can be bent up, joined with spare cable, and bound up with locking wire.

(*c*) A broken chain can be repaired thus: remove sprocket, join chain with cable and bind with locking wire.

(*d*) If a bunch of cables is hit, the cables will in most cases, be severed. In the case of tank cock cables (this will be indicated by the levers falling down under their own weight) it will be possible for the engineer, should the engines not be running on Nos. 2 and 4 tanks, to go behind the front spar of the centre section and operate the cocks by pulling them on with the cables. He should not forget to pull the other tanks off. If the damage has occurred in the wings access to the cable will usually be impossible. Nos. 2 and 4 tanks should, therefore, be turned on immediately.

69. Fire-extinguishers

(i) A graviner fire-extinguisher system is installed in the engine nacelles, each extinguisher being electrically operated by pushbuttons on the port side of the cockpit. Automatic operation is provided by an impact switch in the nose of the aircraft.

(ii) For actions in the event of fire in an engine in flight, *see* A.P. 2095, Pilot's Notes General.

(iii) Hand extinguishers are carried at the following points in the fuselage:

Bomb-aimer's station ..	Under top step between pilots' seats.
Cockpit	Behind starboard pilot's seat.

PART IV—EMERGENCIES

Engineer's station	Beside instrument panel.
W/T operator's station ..	Behind radio equipment framing.
Mid gunner's station ..	On the starboard side forward of the turret.
Tail gunner's station ..	On the port side forward of the turret.

70. **First-aid outfits**

Stowages for three first-aid outfits are provided on the fuselage sides behind the pilots' seats, two on the starboard and one on the port side.

71. **Bomb jettisoning**

In an emergency, the entire bomb load may be jettisoned by means of a pushbutton (24) and a toggle (23) on the right-hand side of the instrument panel. The jettison control is inoperative unless the bomb cell doors are fully open. Bombs are jettisoned by pulling the toggle, and the bomb containers by pressing the pushbuttons; when containers are being carried they should be jettisoned first.

72. **Fuel jettisoning**

(i) The contents of tanks Nos. 2 and 4 on each side can be jettisoned. Cock controls, consisting of four handwheels (95), are mounted on the front spar frame. The wheels are mounted in pairs consisting of a small wheel within a larger one; the large wheels control the No. 4 tank cocks and the small wheels the No. 2 tank cocks. The handwheels are located in the closed position by a small thumb lever in the boss and are rotated anti-clockwise to open the jettison valves. The jettison pipes extend automatically from the underside of the main plane as the valves are opened.

(ii) Before jettisoning, the pilot should set the flaps one-third or more out to prevent fuel flowing over the tail of the aircraft.

(iii) When full, the total quantity of fuel which can be jettisoned from Nos. 2 and 4 tanks is 1,170 gallons—8,500 lb.

PART IV—EMERGENCIES

73. Parachutes

(i) On early aircraft the seats for both pilots, W/T operator and navigator are designed for seat type parachutes. On later aircraft lap-type parachutes are used.

(ii) Parachute (P) and K type dinghy (D)* or combined (C) stowage are installed as follows:

Pilot .. .. P—under forward end of navigator's table.
D—behind port pilot's seat.

Navigator .. .. .. P—on starboard side aft of the front escape hatch.
D—on sloping bulkhead on port side of bomb-aimer's compartment

Flight engineer .. .. C—on aft face of armoured bulkhead on door.

W/T operator .. .. P—on lower part of aft face of armoured door.
D—on forward face of armoured bulkhead, near floor.

Nose gunner .. .. C—on port side of fuselage near turret.

Mid gunner .. .. .. C—on port side of fuselage.

Tail gunner .. .. .. P—on starboard side of fuselage near turret.
D—on port side of fuselage near turret.

Second pilot (when carried) P—on forward face of armoured bulkhead above door.
D—on starboard side behind 2nd pilot's seat.

Extra crew (two) .. .. C—Two; on starboard side opposite door.

* K type dinghy stowages vary on some aircratt.

PART IV—EMERGENCIES

(iii) *Static lines.* These are stowed on the port side of the fuselage, immediately forward of the main fuselage entrance door.

(iv) When incapacitated personnel are being parachuted the aircraft should be flown at about 145 m.p.h. I.A.S. with flaps one-third out. Check that the tail wheels are fully retracted to prevent the parachutes from catching.

74. **Emergency exits**

(i) Parachute exits:

(*a*) Two parachute exits are provided in the floor of the fuselage, one at the rear of the bomb-aimer's station and the other in the centre of the fuselage, just forward of the entrance door; each comprises a hatch hinging inwards. Access to the former exit from the cockpit is by means of two steps between the pilots' seats; a hand lever is provided on the starboard side of these steps to release the catches securing the hatch. On later aircraft the hatch is secured by a catch bar. The rear exit is held by clamp handles.

(*b*) A panel in the starboard side of the fuselage by the tail plane may be pushed out by operating the release bar.

(ii) Dinghy and crash exits:

(*a*) Two escape hatches are provided in the roof, one forward and one aft of the centre section; both hinge inwards and are held by clamp handles. A step is built out from the fuselage side at the front hatch to facilitate exit, and an escape ladder is stowed in the roof at the rear hatch. These escape hatches and the rear entrance door may be opened from the outside by means of keys stowed in adjacent pockets.

(*b*) An outwardly-opening escape hatch is provided in the cockpit roof above the captain's seat and is released by pulling back the catch bar (1) beside the centre roof member.

(*c*) Two knock-out type windows are fitted one on each side of the fuselage aft of the pilots' seats; these are released by delivering a sharp blow towards one edge with the foot or hand.

75. **Dinghy and air/sea rescue equipment**

(i) A type J dinghy—complete with topping up bellows, leak stoppers, rescue line and knife, is stowed in the port wing. It is attached to a point just inside the rear top escape hatch by a steel cable which terminates, at the dinghy end, in a length of 150 lb. breaking-strength cord. The free end of this cord is attached to the dinghy life-line and the knife is supplied for cutting it when ready to cast off. The dinghy may be released and inflated as follows:

On aircraft from which the external manual release has been deleted or blanked off:

(*a*) From inside the fuselage by pulling the handle located on the port side below and just forward of the rear top escape hatch.

(*b*) From outside by a direct pull on the CO_2 head operating cable, reached through the inspection aperture in the dinghy cover in the port wing. The inspection window can be removed after turning two finger-operated fasteners.

(*c*) Automatically by the flooding of the immersion switch located in the fuselage nose.

On aircraft retaining the external manual release:

(*d*) From inside the fuselage by pulling the handle located on the port side, just forward of the bulkhead forward of the rear top escape hatch.

(*e*) From outside by pulling the handle located in a recess just outside the rear top escape hatch, after ripping the fabric patch. OR by a direct pull on the CO_2 operating cable, through the inspection aperture on the dinghy blow-out cover. In this case it is necessary to smash the cellon window.

(*f*) Automatically by the flooding of the immersion switch.

PART IV—EMERGENCIES

(ii) (*a*) On aircraft fitted with large dinghy stowage an emergency pack is provided in the dinghy compartment secured to the dinghy by a coiled lanyard. This contains:

28 tins of water

7 tins of emergency supply rations.

1 Very pistol—1 in. bore.

18 tins of cartridges for above

NOTE.—Aircraft signal pistol cartridges do not fit dinghy pistol

3 fluorescine sea markers

1 first-aid outfit.

1 sponge

2 paddles

1 mast aerial and flag

2 tins of matches

(*b*) On aircraft fitted with the smaller (metal) dinghy stowage a smaller emergency pack is provided. This contains all the items listed above with the exception of the water. On these aircraft a supplementary type 4 and/or type 7 pack is carried in the fuselage on the rear face of the bulkhead on the port side immediately forward of the rear top escape hatch. These contain the water not included in the dinghy stowage pack.

(iii) Single Seat Dinghies.—On later aircraft stowages are provided in the fuselage for K type single seat dinghies in C type packs for each member of crew. *See* para. 73.

(iv) Dinghy Radio Equipment.—(to be transferred to the dinghy after ditching) consists of:

(*a*) Transmitter: stowed on starboard rear face of the wooden bulkhead forward of rear top escape hatch.

PART IV—EMERGENCIES

(*b*) Kite aerial: stowed in starboard side of fuselage roof, adjacent to the rear top escape hatch.

(*c*) Mast aerial complete: Stowed in the emergency pack in the dinghy stowage.

NOTE.—On later aircraft this equipment is stowed with the dinghy.

76. **Ditching**

See A.P. 2095, Pilot's Notes General, and note—best flap seting is one-third out. Flight engineers should see also fuel system management instructions, Part II.

AIR PUBLICATION 1660A, C & D—P.N.
Pilot's and Flight Engineer's Notes

PART V

DATA AND INSTRUCTIONS FOR FLIGHT ENGINEERS

77. **Fuel system**

Details:

(i) Of the seven tanks fitted in each wing, all are self-sealing excepting No. 7. Nos. 2, 4, 5 and 6 are fitted between the spar trusses, Nos. 1 and 3 being behind the rear spar truss. No. 7 tanks are in the leading edges inboard of the inner engine nacelles.

(ii) (*a*) In each wing, Nos. 1, 2 and 7 tanks feed to a distributor on the fuselage side and thence to the inner engines. Nos. 3, 4, 5 and 6 tanks feed to a distributor on the rear spar outboard of the undercarriage bay and thence to the outer engine.

(*b*) The cocks fitted to each tank sump are remotely controlled by levers (89) as are the inter engine balance cocks fitted in the cross feed pipes connecting the two distributors on each side.

(*c*) The two inner distributors are connected by a cross feed pipe, with a direct controlled balance cock (inter system balance cock fitted under a hinged cover in the centre of the floor immediately forward of the rear spar frame).

(iii) An auxiliary tank, consisting of three inter-connected cells, can be carried in the wing bomb cells on each side. These tanks feed to the inner distributors on each side via the remotely controlled auxiliary tank cocks. The controls for these cocks (91) are fitted at the lower corners aft of the spar frame, one on each side.

PART V—DATA FOR FLIGHT ENGINEER

(iv) From each distributor fuel passes through a non-return valve, and filters, to the engine pump and thence through regulating valves to the carburettors to which are fitted the pilot's master cocks (CARBURETTOR COCKS).

Refuelling:

(v) (*a*) Remove small access cover, secured by two cowling clips, and unscrew the filler cap with the appropriate spanner; this is stowed on the starboard side of the fuselage opposite the entrance door.

(*b*) Ensure that the tank cock controls in the fuselage are OFF. Screw the filler extension on to the filler neck (except No. 7 tanks, which are filled direct) and fill tank.

78. Oil system

(i) *For description, see* Part I.

(ii) *Refilling:*

(*a*) Open the access covers over the filler cap and dipstick. These covers are in the nacelle top plating and are clearly marked. The filler cap cover is hinged and has one clip, the dipstick cover has 2 clips.

(*b*) Unscrew the filler cap and knurled cap over the dipstick, using the special spanner stowed with the fuel tank spanner. When filling the tank check contents by dipstick which can be withdrawn by the flanged collar at its upper end.

NOTE.—When the tank is not to be completely filled, add 2 gallons to the amount calculated for the required range to allow for the oil required for propeller operation.

79. Engine details

(i) *Carburettors:*

(*a*) From 0°–65° throttle butterfly opening, corresponding to the fully closed to the first notch position of the throttle levers in the quadrant, main and slow running jets only are in use. This corresponds to the economical cruising range up to +1 lb./sq.in. boost (Hercules XI) +2 lb./sq.in (Hercules VI).

PART V—DATA FOR FLIGHT ENGINEER

(*b*) From 65°–85° throttle opening the main and power jets are in action while beyond 85° throttle opening (at throttle settings in excess of +3½lb./sq.in boost in Hercules XI and + 6 lb./sq.in Hercules VI) the enrichment jet comes into action.

(*c*) Engineers should check on the ground the adjustment of the controls to ensure that the throttle butterfly settings correspond correctly with the positions of the throttle levers.

(*d*) As the carburettor is tuned in S gear a corrector jet is brought into operation when in M gear.

(*e*) Rime or hoarfrost may affect a diffuser tube, air balance screen, and accelerator pump discharge nozzle. Icing should not be experienced with oil temperatures in excess of 60°C. and in any case it practically never occurs on this type of carburettor.

(ii) *Auxiliaries:*

The following auxiliaries are driven by the engines:

Port Inner	*Starboard Inner*
(*a*) Top and front turret hydraulic pump	(*a*) Rear turret hydraulic pump
(*b*) BTH compressor (brakes)	(*b*) 1 RAE compressor (Mk. XIV bomb sight, Auto pilot)
(*c*) 1 vacuum pump	(*c*) 1 vacuum pump
(*d*) 1 24-volt generator	(*d*) 1 24-volt generator

80. **Engine handling**

(i) *Carburettor priming:*

(*a*) If it is necessary to prime the carburettors, *see* Para. 39 (ii), and note the number of strokes quoted therein applies with zwicky type of pump. Some aircraft are fitted with SPE pumps when the following number of strokes should be given if carburettors are known to be empty.

Pump:

SPE Mark I—12 double.

SPE XIXA—6 double.

If engine has not been run for one week or more, half the above number of strokes. If engine has been run within one week, no priming should be necessary.

PART V—DATA FOR FLIGHT ENGINEER

(*b*) Do not rely entirely on fuel pressure gauges or warning lights. Watch the volute casing drains and stop priming *immediately* should fuel drain therefrom.

(ii) *Priming induction system:*
See para. 39

(iii) *Gear changes:*

(*a*) If practicable, request pilot to fly straight and level for approximately five minutes while changing gear.

(*b*) Oil pressure should drop and then rise to the original figure if gear change is operating correctly. If oil pressure is not restored after changing, the gear should be exercised.

(*c*) To prevent a sudden surge of boost when the change from M to S gear is effected, check with the pilot that the boost has been followed back with the throttles before making the change.

(*d*) It is advisable to change gear on the inner engines first, followed by the outer engines rather than change on the two engines on one side first. When changing from M to S gear the control should be operated as smartly as possible to ensure proper operation of the snap-over spring on the engine bulkhead.

(iv) *Cowling gill adjustment:*

(*a*) Always inform the pilot if the cowling gills are open.

(*b*) Provided the maximum cylinder temperatures are not exceeded the gills should be opened as little as possible beyond the " straight " attitude, i.e. in line with the engine cowling, and should never be opened beyond the one-third position which gives the best cooling in flight.

(*c*) When hot air is being used the gills should be fully closed as this increases the efficiency of the hot intake system.

(v) *General:*

(*a*) For maximum efficiency the cylinder temperatures should be between 200 and 220°C.

(*b*) At night the appearance of the exhaust flames gives a good indication of mixture strength, i.e.:

White flame—weak mixture
Red-blue—rich mixture

PART V—DATA FOR FLIGHT ENGINEER

(*c*) If possible avoid low oil temperature. If oil temperature drops below 20° and then increases rapidly it may indicate that the oil in the oil cooler is congealing. The cure is to increase boost and r.p.m. considerably for 20 seconds or so; if this does not cure the trouble the propeller of the engine affected should be feathered when the oil temperature reaches 100°C.

81. **Defects**

(*a*) If a power failure warning light comes on — bright or dim — cut-out is open or fuze has blown; request pilot to increase r.p.m. slightly on engine affected. If light does not go out, check and replace generator fuze; if this does not cure, defect cannot be rectified in flight.
(*b*) On aircraft not fitted with warning lights, a zero reading of a generator voltmeter indicates charge failure; check and replace generator fuze. If voltmeter still reads zero, defect cannot be rectified in flight.

82. **Actions when aircraft is damaged**

(i) *See* Part IV, Para. 68
(ii) *Typical examples:*

Example A

(*a*) Assume that No. 4 (starboard) tank, or the starboard outer engine system has been damaged. This will be indicated by the gauge readings which will show that the outer engine appears to be consuming more fuel than another engine running at the same power.
(*b*) No. 4 starboard tank should then be shut off and any other available tank turned on to supply the starboard outer engine.
(*c*) Should the gauge readings now indicate that the starboard outer engine is still consuming more than another engine running at the same power and that No. 4 starboard tank is not losing fuel, some part of the outer engine pipe system is damaged.
(*d*) The object should now be to use up the fuel left in the whole of the starboard outer system as quickly as possible by running *all* engines from this part of the system, i.e. turn on all balance cocks and turn off all tanks excepting those feeding into the damaged part of the system.

PART V—DATA FOR FLIGHT ENGINEER

(*c*) When this part of the system has been drained the starboard outer engine should be feathered and the starboard outer part of the system isolated by turning off the starboard inter engine balance cock.

Example B.

(*a*) In the example quoted above, should it have been found that No. 4 starboard tank was losing fuel after being shut off, this would indicate that No. 4 tank itself had been damaged. It would then be necessary to use up the fuel remaining in this tank as quickly as possible by turning on all balance cocks and turning off all tanks excepting No. 4 starboard.

(*b*) When this tank has been drained it should be turned off and the balance cocks used to distribute the remaining fuel equally between all four engines.

NOTE.—In the event of damage to the fuel system the aircraft should always be flown at the air speed recommended to give the maximum possible number of air miles per gallon. *See* para. 58.

PART VI

ILLUSTRATIONS

Fig. 1—COCKPIT, GENERAL VIEW

1. Escape hatch release bar.
2. Sliding window.
3. Safety harness.
4. Stowages for portable oxygen bottles.
5. Fire-extinguisher.
6. Windscreen de-icing fluid tank.
7. Propeller speed controls.
8. Folding step (access to bomb aimer's station and front turret).
9. Crash axe.
10. Signal pistol stowage.
11. Armour plating (shown hinged down).
12. Armrest release ring.
13. Armrest (shown hinged up)
14. Instrument panel lamp.

COCKPIT - GENERAL VIEW

Fig. 2

INSTRUMENT PANEL

15. Beam-approach indicator.
16. Cardholder—A.S.I. correction.
17. Undercarriage position indicato .
18. Visual indicator—D.F. loop.
19. Engine-speed indicators.
20. Boost-pressure gauges.
21. D.R. compass switches.
22. Undercarriage master switch.
23. Bomb jettison toggle.
24. Bomb container jettison switch.
25. Duplicate A.S.I.
26. D.R. compass repeater.
27. Bomb door switches and indicator lamps.
28. Brake-pressure gauge.
29. Oxygen flow indicator—2nd pilot.
30. I.F.F. switch.
31. Detonator switches.
32. Booster-coil switches.
33. Rudder pedal adjuster star wheel.
34. Propeller de-icing controls.
35. Undercarriage position indicator switch.
36. Hinged flap over engine-starter switches.
37. Landing lamp switch.
38. Mk.X oxygen regulator.
39. Standard instrument flying panel.
40. Propeller feathering controls.
41. Intercommunication lamp and press button.
42. P.4 compass.
43. Dimmer switch for compass lamp.
44. Card holder—compass deviation.
45. Steering indicator for bomb aiming.

FIG. 2
INSTRUMENT PANEL
INCREASE R.P.M. DECREASE

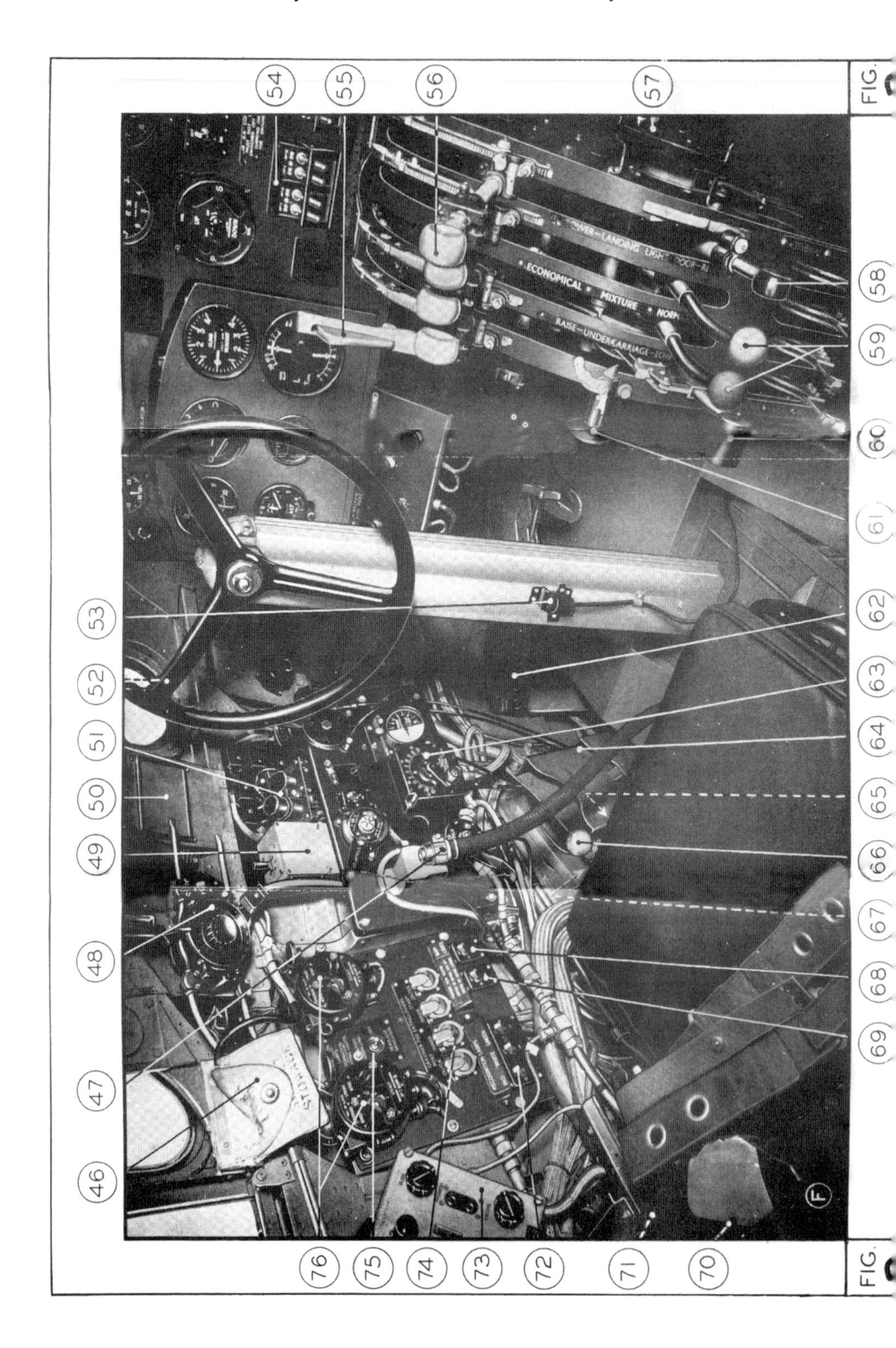
46
47
48
49
50
51
52
53
54
55
56
57
58
59
60
61
62
63
64
65
66
67
68
69
70
71
72
73
74
75
76
FIG.
ECONOMICAL • MIXTURE

Fig. 3

COCKPIT PORT SIDE

46. Stowage for camera control lead.
47. Oxygen conduit clip (port pilot).
48. Remote controls and frequency switch (T.R.9).
49. Mixer box-beam approach.
50. Card holder—D.R. compass correction.
51. Intercommunication socket.
52. Pilot's bomb release switch (location).
53. Connector for (52).
54. Ignition switches.
55. Wheel-brakes lever.
56. Throttle levers.
57. Landing lamp dipping lever.
58. Landing lamp retraction lever.
59. Mixture levers.
60. Undercarriage operating lever.
61. Safety catch—undercarriage operating lever.
62. Indicators for external lamps.
63. Automatic controls panel.
64. Map case.
65. Windscreen de-icing pump.
66. Safety harness release knob.
67. Forced landing flare release toggles.
68. Navigation lamps switch.
69. Pressure-head heating switch.
70. Glove and boot heating socket (location).
71. Seat-height adjusting lever (location).
72. Downward recognition lamps switches.
73. Beam approach control panel.
74. Nacelle fire-extinguisher switches.
75. Headlamp switch.
76. Signalling switch boxes.

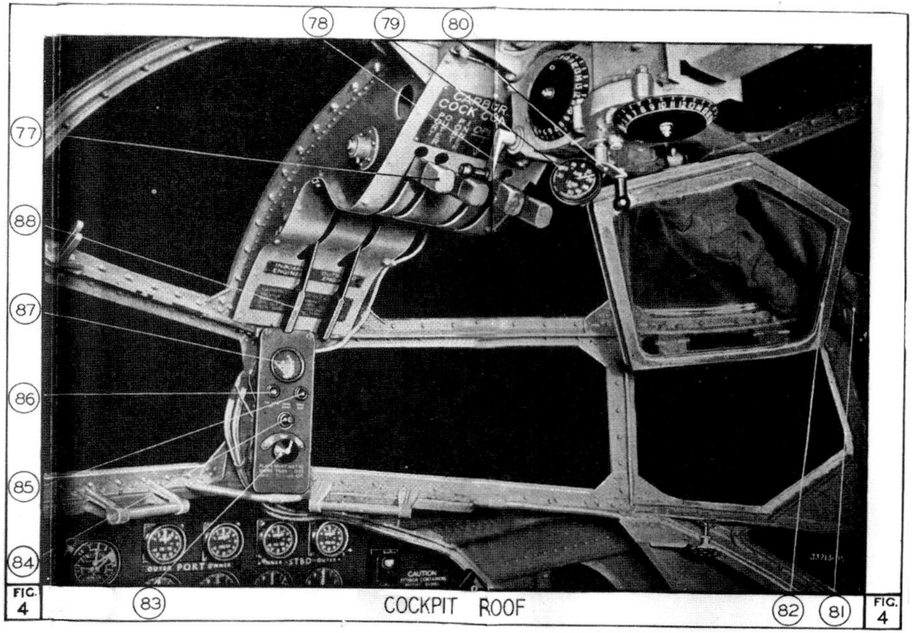

Fig. 4

COCKPIT ROOF

77. Pilot's master cock controls.
78. Elevator trimming tab handle.
79. Air temperature thermometer.
80. Rudder trimming tab handle.
81. Window blind.
82. Hinged window panel.
83. Flaps control knob.
84. Flaps position-indicator switch.
85. Flaps-indicator warning lamp.
86. Flaps one-third-out warning lamp.
87. Flaps-position indicator.
88. Slow-running cut-out controls.

FIG. 5 FLIGHT ENGINEER'S CONTROLS — PORT SIDE

KEY TO *Figs. 5 and 6*

Fuel tank cock controls (1 to 6) ..	89	Supercharger controls	93
Carburettor air-intake controls ..	90	Cabin heating controls	94
No. 7 tank cock controls	91	Fuel jettison valve controls	95
Trailing aerial winch	92	Electrical distribution panel ..	96

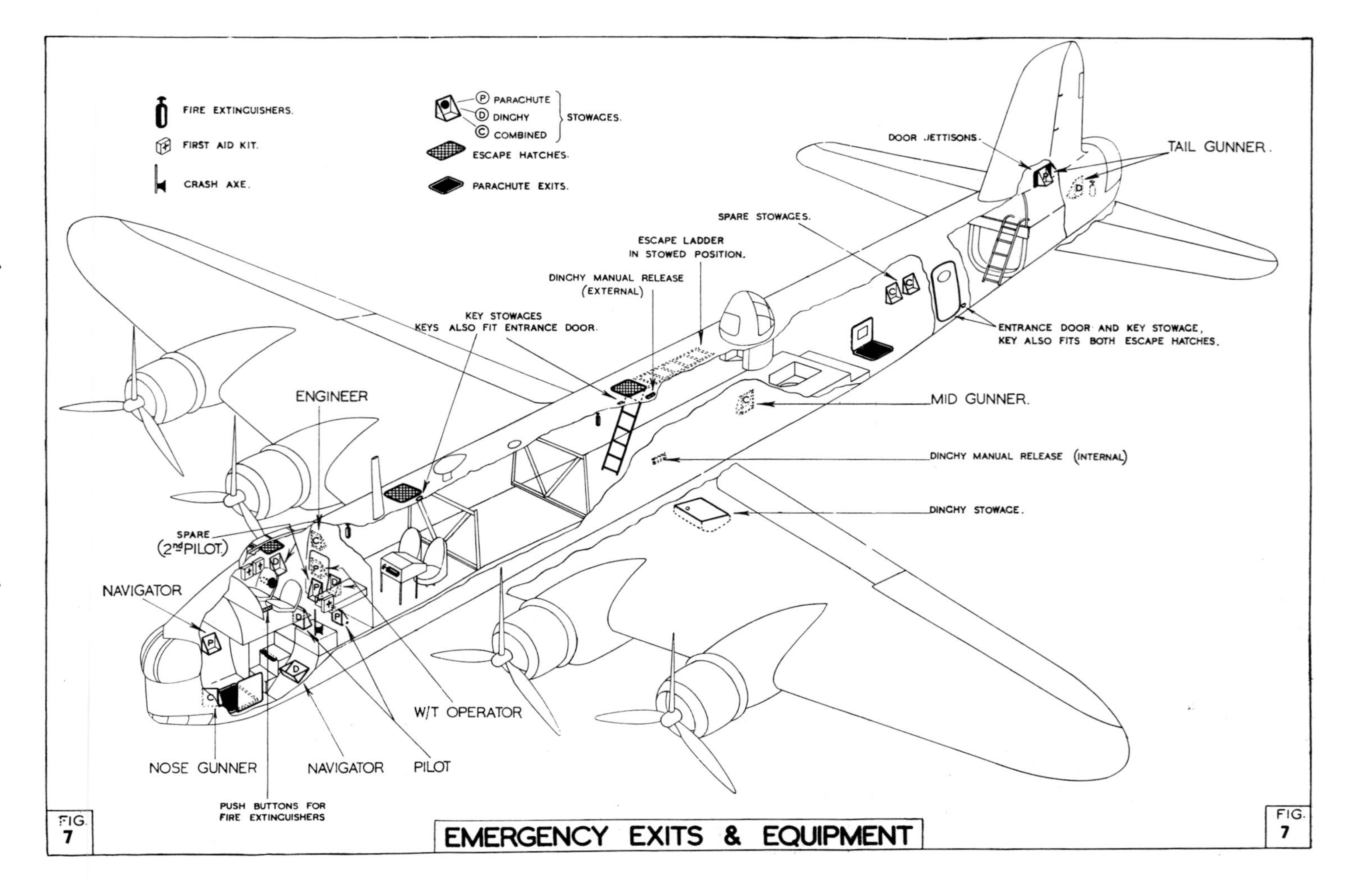
FIRE EXTINGUISHERS.
FIRST AID KIT.
CRASH AXE.
(P) PARACHUTE
(D) DINGHY
(C) COMBINED
STOWAGES.
ESCAPE HATCHES.
PARACHUTE EXITS.
DOOR JETTISONS.
TAIL GUNNER.
SPARE STOWAGES.
ESCAPE LADDER IN STOWED POSITION.
DINGHY MANUAL RELEASE (EXTERNAL)
KEY STOWAGES
KEYS ALSO FIT ENTRANCE DOOR.
ENTRANCE DOOR AND KEY STOWAGE,
KEY ALSO FITS BOTH ESCAPE HATCHES.
ENGINEER
MID GUNNER.
DINGHY MANUAL RELEASE (INTERNAL)
DINGHY STOWAGE.
SPARE (2nd PILOT)
NAVIGATOR
W/T OPERATOR
NOSE GUNNER
NAVIGATOR
PILOT
PUSH BUTTONS FOR FIRE EXTINGUISHERS
FIG. 7
EMERGENCY EXITS & EQUIPMENT
FIG. 7

HANDLEY PAGE HALIFAX

HALIFAX III

NOTES TO USERS

THIS publication is divided into six parts: Descriptive, Handling Instructions, Operating Data, Emergencies, Additional Data for Flight Engineer, and Location of Controls and Illustrations. Part I gives only a brief description of the controls with which the pilot and flight engineer should be acquainted.

These Notes are complementary to A.P. 2095 Pilot's Notes General and assume a thorough knowledge of its contents. All pilots should be in possession of a copy of A.P. 2095 (*see* A.M.O. A93/43). Flight engineers should also have a copy of A.P. 2764 to be issued shortly in provisional form.

Words in capital letters indicate the actual markings on the controls concerned.

Additional copies may be obtained from A.P.F.S., Fulham Road, S.W.3, by application on R.A.F. Form 294A, in duplicate, quoting the number of this publication in full—A.P. 1719C & G—P.N.

Comments and suggestions should be forwarded through the usual channels to the Air Ministry (D.T.F.).

AIR MINISTRY
March 1944
(*Reprinted: October* 1944)

AIR PUBLICATION 1719C & G—P.N.
Pilot's and Flight Engineer's Notes

PILOT'S & FLIGHT ENGINEER'S NOTES HALIFAX III

LIST OF CONTENTS

PART I—DESCRIPTIVE

OPERATIONAL CONTROLS AND EQUIPMENT

PART II—HANDLING INSTRUCTIONS

PART III—OPERATING DATA

PART IV—EMERGENCIES

PART V—ADDITIONAL DATA FOR FLIGHT ENGINEER

PART VI—LOCATION OF CONTROLS AND ILLUSTRATIONS

AIR PUBLICATION 1719C & G—P.N.
Pilot's and Flight Engineer's Notes

PART I

DESCRIPTIVE

INTRODUCTION

The Halifax III aircraft is a heavy bomber with four Hercules VI or XVI engines and de Havilland hydromatic fully-feathering propellers.

FUEL AND OIL SYSTEMS

Mark III only

1. **Fuel Tanks.**—Fuel is carried in twelve self-sealing tanks.

The capacities are:

2—No. 1 inboard tanks	..	247 galls. each
2—No. 2 inner-wing nose tanks	..	62 ,,
2—No. 3 centre tanks	..	188 ,,
2—No. 4 outboard tanks	..	161 ,,
2—No. 5 outer engine tanks	..	122 ,,
2—No. 6 outer engine tanks	..	123 ,,

Each No. 5 tank is connected to the adjacent No. 6 tank so that tanks Nos. 5 and 6 together may be considered as one tank containing 245 gallons of fuel. Provision is made for the following auxiliary tanks:

(i) One, two or three self-sealing tanks of 230 gallons each, carried in the fuselage bomb bay.

(ii) Two self-sealing tanks of 96 gallons each, carried in the outer bomb compartments, one in each wing.

(iii) Two tanks for exceptional long-range operations may be fitted port and starboard on the rest seats in the centre fuselage.

A nitrogen fire protection system for all tanks is to be installed; nitrogen is fed into the tanks automatically as fuel is used, so that no inflammable petrol-air mixture is present in the tanks. The control valve is on the port side of the fuselage at the rest station and must be fully opened before any petrol is used. No further attention is necessary.

PART I—DESCRIPTIVE

2. **Fuel cocks**

(i) The supply to each engine can be shut off by the master engine cock controls which are mounted on the bulkhead aft of the first pilot and accessible to his right hand. In each case, the cock is moved to the ON position by pushing the lever up. These fuel cocks also operate the slow-running cut-outs for stopping the engines when moved to the OFF position.

(ii) Each tank has its own ON-OFF cock (tanks Nos. 5 and 6 being considered as one tank), the cocks being under the control of the flight engineer. The cock-control levers are fitted on the forward end of the rest seats.

(iii) The fuel systems in each wing for port and starboard engines are identical and entirely independent, but are interconnected by a cross-feed pipe and cock. This cock, which is normally kept shut, is on the aft face of the rear spar, under the control of the flight engineer. Port and starboard cross-feed cocks are provided in each fuel system to segregate the sets of tanks supplying individual engines; thus with these cocks closed, tanks Nos. 1 and 2 supply inboard engines and tanks Nos. 3, 4, 5 and 6 supply outboard engines. It will be noted that the total capacity of tanks Nos. 1 and 2 is less than that of tanks Nos. 3, 4, 5 and 6, and thus if all fuel is to be consumed, a time arrives when the inboard engine must draw fuel from the outboard group of tanks. All cross-feed cocks should be kept closed at take-off and when the aircraft is over the target. The port and starboard cross-feed cock controls are mounted on the forward end of the rest seats, and the cocks should be opened only when it is necessary to feed two engines from one tank.

(iv) The distributing cocks for the three fuselage bomb-bay tanks are mounted under the step aft of the front spar. There are no cocks for the wing bomb-bay tanks, but where Mod. 981 has not been embodied, the fuselage bomb-bay distributing cocks must be set to No. 3 tanks before transferring fuel from the wing bomb-bay tanks.

3. **Immersed pumps.**—An immersed pump is fitted in each of the five auxiliary tanks. The pumps in the fuselage tanks transfer fuel to tanks Nos. 1 or 3, whichever is

PART I—DESCRIPTIVE

selected, and the pumps in the wing bomb-bay tanks transfer fuel direct into tank No. 1. The switches to the pumps are under the control of the flight engineer.

4. **Fuel contents gauges.**—Gauges for tanks Nos. 1, 2, 3, 4, 5 and 6 (and the wing long-range tanks if fitted) are mounted on the engineer's panel. The contents gauges are direct reading and a circuit switch is mounted beside them. The fuel pressure warning lamps are mounted on the engineer's panel and duplicated at the fuel cock controls on the rest seats.
The fuselage bomb-bay tanks are fitted with direct reading contents gauges visible through the bomb-bay hatches in the fuselage floor.

5. **Fuel priming pumps.**—A priming pump together with carburettor and cylinder priming cocks is mounted on a panel at the rear of each undercarriage fairing and serves both engines on one side.

6. **Oil system**

(i) Two self-sealing oil tanks mounted in the nose portion of the intermediate wing on both sides serve the inner and outer engines. The outboard tanks contain 32 galls. of oil with 10 galls. air space, and the inboard tanks 39 galls. of oil with 5 galls. air space.

(ii) Four oil dilution switches (62) are fitted, adjacent to the engine starting switches (55) on the engineer's panel.

MAIN SERVICES

7. **Hydraulic system (Messier)**

(i) A Lockheed pump fitted on the starboard inner engine supplies power to feed the following services:

Undercarriage (and tailwheel on later aircraft)
Flaps
Bomb doors
Air-intake
Landing lamps

A feature of the undercarriage, flaps and bomb-doors circuits is the use of the pump to operate the jacks in one direction only (i.e. to raise the undercarriage and flaps

PART I—DESCRIPTIVE

and to close the bomb doors); the fluid above the piston is thereby forced into an accumulator, increasing the air pressure in the accumulator and thus storing energy which is used to operate the jack in the reverse direction when required. The air-intake and landing lamp circuits differ from the main circuits in that the pump operates the jacks in both directions.

IMPORTANT NOTE.—The undercarriage, flaps or bomb door levers, after having been operated, should be returned to the " neutral " position thereby closing the distributor. Should a lever be left in its operating position and the pipe lines from the distributor to the jacks be damaged, fluid would be automatically discharged through the damaged pipe thus leaving no fluid for the operation of the remaining undamaged circuits. The undercarriage lever should, however, be left in the " down " position at all times when the undercarriage is down, and the bomb doors lever should be operated on the descent as described in para. 28.

(ii) An hydraulic handpump is provided on the front spar on the port side of the fuselage, and should be used to operate the system through the normal pipe-lines and controls, when the engine-driven pump is not working.

(iii) Emergency circuits operated by the engineer are provided to lower the undercarriage and open the bomb doors. Control is effected by means of emergency cocks mounted on the front face of the front spar, and these cocks when opened, allow the engine-driven pump or handpump to perform the functions normally carried out by the accumulators.

8. **Pneumatic system.**—A Heywood compressor for operating the wheel brakes, and an R.A.E. compressor for operating the automatic pilot are mounted on the port inner engine. The R.A.E. compressor also works the computor for the Mark XIV bomb sight so that if the bomb sight is in use the gyro of the automatic pilot must be set to OUT. A Pesco vacuum pump on the port inner engine supplies the instrument flying panel, and

PART I—DESCRIPTIVE

another Pesco pump on the starboard inner engine operates the Mark XIV bomb sight. A suction gauge (10) and a change-over cock (16) marked NORMAL and EMERGENCY is mounted on the instrument panel, and the starboard pump can be used for the instrument flying panel by setting the change-over cock to EMERGENCY, but in this event the Mark XIV bomb sight cannot be used.

9. **Electrical system.**—Three 1,500-watt generators, driven by the port outer and both inner engines are connected in parallel and feed two 24-volt, 40-amp. hr. batteries for supplying the usual electrical services, cowling gill motors and dorsal and rear gun turrets. An AC generator driven by the starboard outer engine supplies current for the special radio installations. A GROUND/FLIGHT switch is fitted on the starboard side of the fuselage at the flight engineer's station.

AIRCRAFT CONTROLS

10. **Flying controls**

(i) The dual controls are not a permanent fitment but conversion sets are supplied and fitted if required.

(ii) The rudder pedals can be adjusted on the ground by removing a nut and bolt from the shank of the pedal, sliding the pedal backwards or forwards as required, and refitting the bolt which should then be locked by a split pin. Finer adjustment can be made by a foot-operated centrally placed starwheel (33).

11. **Locking of flying controls** (*see* page 11).—Before attaching the locking gear, the controls should be moved to the neutral position. The aileron control should be locked by applying the special locking plate to the handwheel and securing it to the block on the column with the bolt provided.

The plate carries a tube which crosses the pilot's seat when the plate is in position, and prevents the seat from being occupied before the controls are unlocked. The rudder and elevator controls should be locked with the locking lever connected to the elevator and rudder control

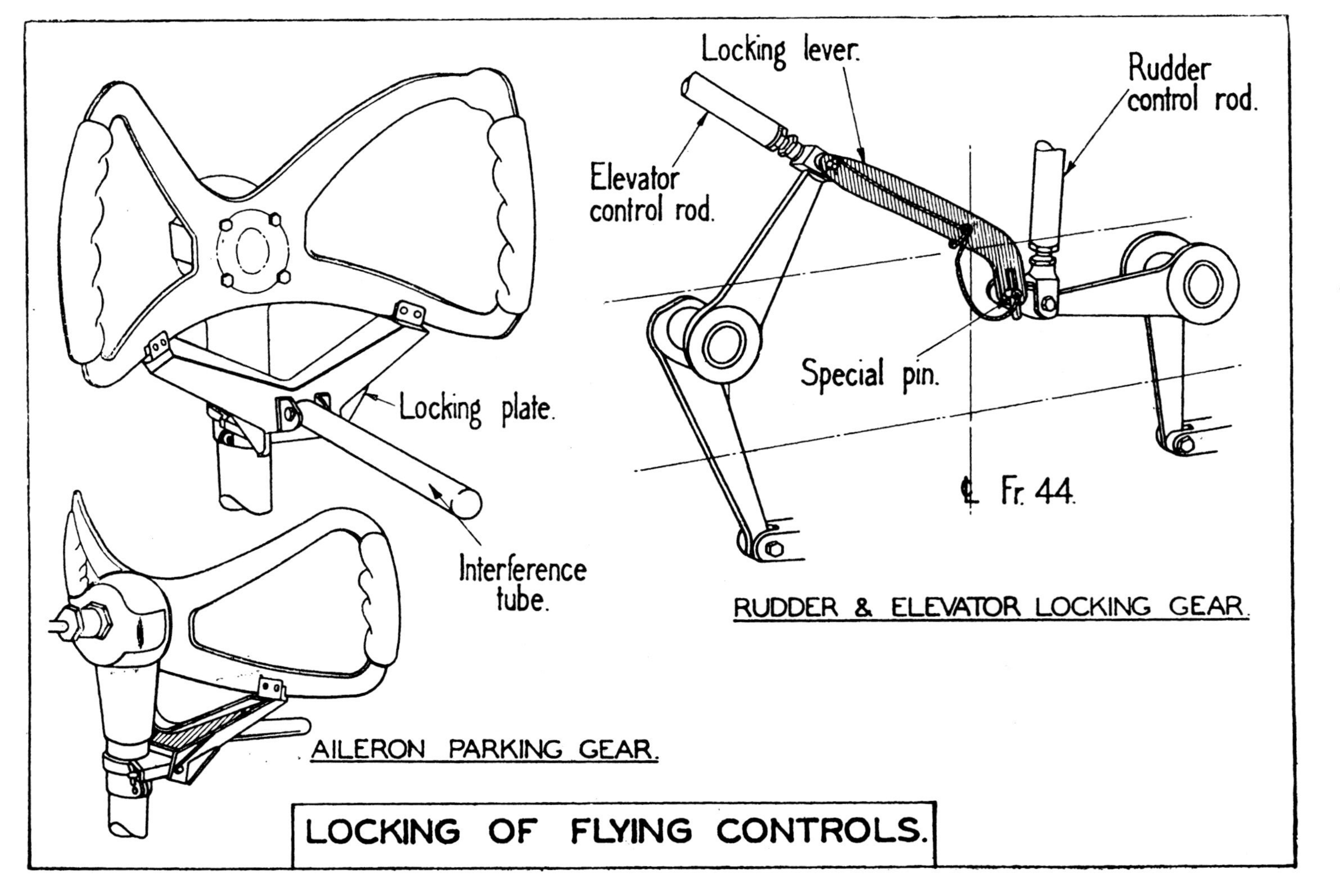
Locking lever.
Rudder control rod.
Elevator control rod.
Special pin.
℄ Fr. 44.
Locking plate.
Interference tube.
RUDDER & ELEVATOR LOCKING GEAR.
AILERON PARKING GEAR.
LOCKING OF FLYING CONTROLS.

PART I—DESCRIPTIVE

rods at the tail. The locking lever is fixed at its top end on a spigot on the elevator control and held in position by a fastener spring attached by lanyard to the locking lever. At its lower end the locking lever is secured to the rudder control and the adjacent fuselage member by means of a special screwed pin also attached by lanyard to the locking lever.

A.L.3
Part I
Para.
12 (i)

12. **Trimming tabs**

(i) The trimming tab controls (77 & 79) for the starboard aileron and rudder are on the left-hand side of the cockpit and work in the natural sense. These controls have been raised a few inches on Mark VII aircraft and are more conveniently placed.

(ii) The elevator trimming tab handwheel (32) is mounted on the centre line of the aircraft accessible to both pilots. An indicator scale (51) along which a pointer travels is mounted aft of the wheel. The movement of the wheel is also in the natural sense.

13. **Automatic pilot.**—The controls include main switch (68), clutch lever (78), control cock (76), attitude control (75), and steering control (69) mounted on the port side of the cockpit. A combined pressure and trim gauge (28) is on the lower port side of the instrument panel. (For operating instructions *see* A.P. 2095, Pilot's Notes General.)

A.L.2
Part I
Para. 14

14. **Undercarriage controls.**—The hydraulic operation of the undercarriage is controlled by the longest lever (29) mounted immediately below the flaps and bomb doors control levers on the right of the pilot's seat. A " neutral " position is provided and after raising the undercarriage, the lever should be returned to this position. For raising the undercarriage the lever is pulled up. The flight engineer has mechanical up-locks under his control, one on each side of the fuselage above the rest seats. These must be engaged at all times when the undercarriage is retracted except during local flying and must be pulled out before the undercarriage is lowered. The control should be left in the " down " position after lowering the undercarriage. An electrical interlocking mechanism is provided to prevent the undercarriage being retracted until the weight of the aircraft is off the wheels. This mechanism consists of a solenoid which becomes energised when the weight of the

PART I—DESCRIPTIVE

aircraft is off the wheels, and withdraws a bolt which then allows the undercarriage lever to be moved through the " neutral " position. Should it ever become necessary to override this safety device, hold back a small lever located through a hole in the right-hand side of the control box.

15. **Undercarriage indicators.**—A single instrument (23) on the pilot's sloping panel indicates whether the undercarriage or tailwheel is unlocked or locked in the down position. A control knob in the centre of the instrumen enables the lamps to be dimmed for night flying and may also be used to bring a duplicate set of lamps into operation in the event of breakage.

The operation of the instrument is as follows:

Upper lamps—red	Unlocked positions
Lower lamps—green	Locked down position
No lamps	Undercarriage up and mechanical up-locks engaged

16. **Undercarriage warning horn.**—A horn on the port side of the cockpit sounds to warn the pilot when any throttle is closed two-thirds or more and the undercarriage is not locked down.

17. **Flaps control.**—The control (30) for operating the flaps is the shorter (starboard) of the two small levers on the hydraulic control box mounted on the right of the pilot's seat. The movement of the lever is in the same sense as the flaps, i.e. the upward movement of the lever raises the flaps.

NOTE.—The flaps lever should be returned to " neutral " after each operation.

In emergency the flaps can be raised by use of the hand-pump. An isolating cock is fitted between the flaps hydraulic accumulator and the jacks, and is closed by the flight engineer when the flaps are not in use. If this isolating cock is not closed and one of the hydraulic pipe-lines is damaged, the flaps will immediately lower under the accumulator power which is thereby released.

PART I—DESCRIPTIVE

18. **Flaps position indicator.**—An electrical indicator (4) is mounted on the lower portion of the pilot's instrument panel immediately forward of the first pilot's control column. It shows the settings of the flaps by means of a pointer which moves over a scale graduated in degrees and marked UP and DOWN at the extremities.

19. **Brakes control.**—The brake levers (35) are mounted on the first pilot's handwheel, on all aircraft, and are duplicated on the second pilot's handwheel on certain aircraft when dual control sets are fitted. There is a parking catch on the underside of the control. The brakes are also operated differentially by the rudder bar. A triple pressure gauge (19) is mounted on the starboard side of the instrument panel.

ENGINE CONTROLS

20. **Throttle controls.**—Three positions are marked in luminous paint on the throttle quadrants, viz. ECB (weak mixture cruising boost) and RB (climbing boost), and another line indicating the mid-position between these two settings. The rear faces of the throttle levers are also marked with a luminous line. To prevent the throttle levers moving under engine vibration, a friction lever (18) is fitted on the starboard side of the control box. This lever locks the throttle levers in any desired position when pulled aft.

21. **Mixture control**

(i) Where Hercules VI engines are fitted, a lever mounted in the centre of the bottom tier of the engine control box controls the mixture. The WEAK setting is the upper position and the RICH setting the lower position.

(ii) Where Hercules XVI engines are fitted the lever is inoperative and should be wired. Weak mixture is obtained with the throttle lever up to the ECB position and rich mixture with throttle lever more than 4° forward of this. Excessive cylinder temperatures may occur if the positions between these are used. When climbing

PART I—DESCRIPTIVE

at high altitudes, the mixture strength at the RB setting is too rich, resulting in loss of power. The position midway between RB and ECB gives a mixture suitable when boost has fallen to + 3 lb./sq.in.

22. **Propeller speed controls.**—The four propeller speed control levers (46) are mounted in the centre of the control box. The control levers are moved upwards to increase, and downwards to decrease the engine revolutions. A friction lever similar to the one described in para. 20 is fitted on the starboard side of the control box.

23. **Two-speed supercharger controls.**—A single lever (49) for the four superchargers is situated on the starboard side of the mixture control lever in the bottom tier of the control box. The upper position of the lever is for HIGH gear and the lower position for LOW gear.

24. **Carburettor air-intake heat control.**—The control lever is on the port side of the mixture control lever in the bottom tier of the engine control box. The lever has two positions marked HOT AIR and COLD AIR. The COLD AIR position should always be used unless the intake becomes iced up.

25. **Cowling gill controls.**—The gills are controlled electrically, the gill motor switches, warning lights which show when the gill motors are operating, and the gills indicators are mounted together on the flight engineer's lower instrument panel.

26. **Slow-running cut-outs.**—These are operated by closing the pilot's master engine cocks.

27. **Ignition and starting controls.**—The ignition switches are on the pilot's instrument panel; the booster coil and engine starting pushbuttons are on the flight engineer's panel.

OPERATIONAL CONTROLS AND EQUIPMENT

28. **Bomb doors control.**—The control (31) for operating the bomb doors is the longer (port) of the two levers on the hydraulic control box mounted between the two pilot's seats aft of the elevator trimming tab controls.

PART I—DESCRIPTIVE

It is beside the lever operating the flaps but works in the opposite sense; when the lever is moved to the up position the bomb doors open. A "neutral" position is provided for the lever, and after the bomb doors have been opened or closed the lever must be returned to this position. An isolating cock is fitted between the bomb doors accumulator and the jacks. If this isolating cock is not closed and the hydraulic pipe-lines are damaged, the bomb doors will immediately open under the accumulator power which is thereby released. However, the isolating cock should be left open on the outward journey of an operational flight, and should only be closed after the bombs have been dropped and the bomb doors closed; otherwise the need to open the isolating cock may cause delay when jettisoning bombs in an emergency. In addition, a selective-closing valve (located on the rear of the pilot's bulkhead) also operated by the flight engineer, is incorporated in the bomb doors circuit. When this valve is closed the bomb doors may be opened in the normal manner, but the fuselage bomb doors cannot be closed again. The valve is normally left open and is only closed when large bombs are carried, which necessitate the fuselage bomb doors remaining partially open. In the latter case the bomb doors are partially closed by using the handpump before the engines are started, and then the selective-closing valve is closed and ***must not*** be opened until after the large bombs have been released.

NOTE.—On aircraft where Mod. 1085 is not incorporated, in order to relieve the pressure built up due to thermal expansion of hydraulic fluid on descending to warmer air after operating bomb doors over the target, the bomb doors lever should be momentarily operated as follows for every 15° rise in temperature.

(*a*) Select bomb doors CLOSED.
(*b*) Select bomb doors OPEN.
(*c*) Select bomb doors CLOSED.
(*d*) Return to neutral.

PART. I—DESCRIPTIVE

29. **Bomb release controls.**—The bomb jettison control (14) and bomb door warning lights (15) are mounted on the starboard side of the main instrument panel, and the pilot's bomb firing switch (21) on the sloping instrument panel below it. The bomb release is inoperative until the bomb doors are open, and the bomb door warning lights are duplicated at the bomb-aimer's station.
The pilot is not provided with a selector switchbox, but he may fire any bombs selected by the bomb-aimer by pressing the bomb firing switch button (21) on the sloping instrument panel. Before the bomb doors are opened the trailing aerial must be wound in by the W/T operator. (For details of bomb jettisoning *see* para. 61.)

30. **Camera control.**—A camera control switch is mounted in the pilot's cockpit just forward of the rudder tab control handwheel and is operable only when the bomb-aimer's camera control is fully connected.

31. **Paratroop signalling switch.**—A switch and two indicator lamps (17) for paratroop release are mounted on the starboard side of the pilot's instrument panel. A similar switch and lights are duplicated at the bomb-aimer's station and also at the paratroop station in the rear fuselage.

32. **Landing lamp control.**—The selector lever (20) for operating the landing lamp is mounted on the left-hand side of the throttle control box. The lever is normally in a central position and must be held in the down or up position until the landing lamp reaches the desired angle. The hydraulic circuit is operated in both directions by the engine-driven pump, but if necessary can be operated by the handpump.

AIR PUBLICATION 1719C & G—N.P.
Pilot's and Flight Engineer's Notes

PART II

HANDLING INSTRUCTIONS

NOTE.—All speeds quoted are for aircraft with the static side of the pilot's A.S.I. connected to the pressure head. Speeds given in brackets are for aircraft with the static vent on the port side of the rear fuselage connected; in which case the letters SV will be painted on the instrument panel adjacent to the A.S.I.

33. **Management of fuel system** (Mark III only. For Mark VII *see* Fig. 7).

In using the fuel system, there are several methods which may be adopted, but all methods should conform to the following requirements and limitations:

(*a*) The fuel tanks should be used in such an order as to require a minimum of cock changes, so arranged that even when full fuel is not carried, the sequence of cock changes remains as far as possible the same.

(*b*) The supply of fuel should be arranged so that at no time will two engines on the one side of the aircraft cut together through lack of fuel. If a tank is allowed to run dry, it will then suck air and cause the engine to cut. Should the wing crossfeed cock be open, the other engine will also cut.

(*c*) All crossfeed cocks should remain closed at take-off and also when the aircraft is over the target. This is necessary in case part of the system should be damaged; if the cocks were open, all the engines might cut.

(*d*) No engine should be run at any time on two tanks simultaneously, and when a tank empties, its cock should be turned off before another tank is turned on. As the fuel pressure warning lights are not visible to the pilot,

PART II—HANDLING INSTRUCTIONS

the flight engineer must inform him as soon as the lights show so that the relevant engine(s) may be throttled back before the tank change is made : re-open throttle(s) slowly after change-over. A tank should never be drained when flying below 3,000 feet, and tank changes should as far as possible be avoided below this height.

(*e*) When long-range tanks are carried in the bomb bays, their contents should be used as early in the flight as possible, in order that the flight duration can be reassessed in the event of any of the fuel transfer pumps failing.

(*f*) No. 2 tanks should not be used for take-off or landing.

(*g*) It is necessary for strength considerations of the aircraft structure that if the aircraft is to take-off with reduced fuel load at weights exceeding 55,000 lb., tanks Nos. 5 and 6 should be filled rather than the tanks further inboard.

(*h*) Specimen schedule of tank changes

Six wing tanks—1,806 galls.

	Tank contents at change					
	1	2	Cross-feed cock	3	4	5 & 6
Take-off on 1 and 3	247	62	closed	188	161	245
Use 80 galls. Change to 5 & 6	167	62	open	108	161	245
Use 220 galls. Change to 1 and 4	167	62	close	108	161	25
Use 60 galls. Turn off 1	107	62	open	108	101	25
Use 50 galls. Change to 2 and 5 & 6	107	62	close	108	51	25
Drain 5 & 6. Turn on 4	107	37	closed	108	51	0
Drain 2. Change to 1 and 3	107	0	closed	108	14	0

PART II—HANDLING INSTRUCTIONS

Six wing tanks and one fuselage tank—2,036 galls.

	Tank contents at change						
	Fuse-lage	1	2	Cross-feed cock	3	4	5 & 6
Take-off on 1 and 3	230	247	62	closed	188	161	245
Use 80 galls. Change to 5 & 6	230	167	62	open	108	161	245
Pump fuselage tank into 1 until full and remainder into 3 while using 220 galls. from 5 & 6. Change to 1 and 4	0	247	62	close	143	161	25
Use 120 galls. Change to 2 and 5 & 6	0	127	62	closed	143	41	25
Drain 5 & 6. Change to 4	0	127	37	closed	143	41	0
Drain 2. Change to 1 Drain 4. Change to 3	0	86	0	closed	102	0	0

34. **Preliminaries**

(i) Before entering the aircraft, check:
Engine, cockpit and pitot head covers off
Visual check for oil and fuel leaks
Cowlings secure
Tyres for cuts and creep; oleo legs for even compression
Undercarriage accumulator pressure (250 lb./sq.in. minimum)

(ii) On entering the aircraft, check:
All loose equipment stowed
Turrets central and engaged
All controls unlocked; locking gear stowed
GROUND/FLIGHT switch to FLIGHT
Indicators and lights
Tailwheel accumulator pressure (250 lb./sq.in.)
Flaps accumulator pressure (1050 lb./sq.in. flaps up)
" " " (700 lb./sq.in. flaps down)
Flaps isolating cock unscrewed
Up-locks disengaged and clips secured

PART II—HANDLING INSTRUCTIONS

Bomb-doors accumulator pressure (600 lb./sq.in. doors closed)
,, ,, ,, (400 lb./sq.in. doors open)
All crossfeed cocks OFF
Fuel contents
Nitrogen valve on (if fitted)

(iii) Before starting engines, check:
Flaps up and landing lamp retracted
Undercarriage lever down, and flaps and bomb door levers neutral
Brake pressure: brakes on
Flying controls
Oxygen capacity and flow
Test visual call light system

35. **Starting engines and warming up**

(i) The engine starting and booster coil switches, gills, temperature gauges, pressure gauges and indicators are under the charge of the flight engineer, but the pilot should be in his seat to see that the following sequence of actions is carried out. The engines should be started in turn; an engine should not be primed until its turn for starting comes.

(ii) Have ground battery plugged in and GROUND/FLIGHT switch turned to GROUND: turn on master engine cocks and instruct flight engineer to turn ON tanks 1 and 3.

Set engine controls as follows:

Throttles	Just off rear stops
Mixture control (if fitted) ..	Down
Propeller	Fully up
Superchargers	M ratio
Air-intake heat control ..	COLD AIR
Gills	OPEN

(iii) Have each engine turned slowly by hand for at least two revolutions, to avoid the danger of hydraulicing.

(iv) Switch on the ignition, and press the starter and booster coil buttons. Turning periods must not exceed 20 seconds with a 30 seconds wait between each. The ground crew will work the priming pump while the engine is being

PART II—HANDLING INSTRUCTIONS

A.L.3 Part II Para. 35(iv)

turned: it should start after the following number of strokes if cold:

Air temperature °C.	+30	+20	+10	0	—10	—20
Normal fuel	1	1	2	3		
H.V. fuel (Mark VII)				1	2	4

It will probably be necessary to continue priming after the engine has fired and until it picks up on the carburettor.

(v) When the engine is running satisfactorily, remove finger from the booster coil button. The ground crew will screw down the priming pump and turn off the priming cocks.

(vi) Open up each engine gradually to 1,000 r.p.m. and warm up at this speed.

(vii) DR compass to ON and SETTING.

(viii) Ground battery disconnected: GROUND/FLIGHT switch to FLIGHT.

36. **Testing engines and installations**

While warming up:

(i) Check temperatures and pressures, and test operation of hydraulic system by lowering and raising the flaps. Test each magneto as a precautionary check.

After warming up, and for each engine in turn:

NOTE.—The following comprehensive checks should be carried out after repair, inspection (other than daily), or at the pilot's discretion. Normally they may be reduced in accordance with local instructions.

(ii) Open up to 1,500 r.p.m.; exercise and check operation of two-speed supercharger. Oil pressure should drop momentarily at each change and r.p.m. should fall when S ratio is engaged. Return control to M ratio.

(iii) At +2 lb./sq.in. boost, check operation of constant-speed propeller. Return lever fully up.

(iv) Open throttle and check take-off boost and static r.p.m., which should be 2,800 at take-off boost. Throttle back until boost is below +6 lb./sq.in. and a slight drop in r.p.m. is noted, then test magnetos by switching off each in turn. The additional drop in r.p.m. must not exceed 50.

PART II—HANDLING INSTRUCTIONS

NOTE.—It may be found that, although the drop in r.p.m. is hardly perceptible, undue rough running is experienced, in which case the ignition system should be checked.

37. Check list before taxying

Wheel brakes: Brake pressure (90 lb./sq.in.)
Supply pressure (200 lb./sq.in. minimum)

DR compass switches to ON and NORMAL

Auto-controls: Switch	..	OFF
Clutch	..	IN
Gyro	..	OUT
Instrument flying panel	..	Check vacuum on each pump
Pitot-head heater switch	..	ON

38. Check list before take-off

T—Trimming tabs	..	Elevator: 2 divisions tail-heavy Rudder: neutral Aileron: neutral
M—Mixture control (if fitted)		Down
P—Propeller speed control		Fully up
F—Fuel	..	Engineer checks cock settings
F—Flaps	..	0°–35° down
Superchargers	..	M ratio
Gills	..	⅓ open
Air-intake ..	..	COLD AIR

A.L.2 Part II Para. 39 (i)

39. Take-off

(i) Open up to –2 lb./sq.in. boost against the brakes to see that the engines are responding evenly, then throttle back and release the brakes. Open the throttles slowly at first, then fully as the aircraft accelerates. There is a slight tendency to swing to starboard but the aircraft can be kept straight initially on the throttles, and, as the speed increases, by the rudders.

PART II—HANDLING INSTRUCTIONS

(ii) The tail comes up easily as speed develops. The forward force required on the control column is not large.

(iii) At 55,000 lb. the aircraft can be pulled off the ground at 95 (105) m.p.h. I.A.S.; at 63,000 lb. it can be pulled off at 100 (110) m.p.h. I.A.S.

(iv) Safety speed is 145 (155) m.p.h. I.A.S.

A.L.2 Part II Para. 40 (i)

40. Climbing

(i) The initial speed for maximum rate of climb is 130 (140) m.p.h. I.A.S. at which speed gills must be fully open, but since this is below safety speed, a climbing speed of 145–150 (155–160) m.p.h. I.A.S. with gills closed is recommended.

(ii) After the undercarriage and flaps have been raised, the flight engineer should close the flap isolating cock and engage the undercarriage mechanical up-locks.

41. General flying

(i) *Stability.*—The aircraft is stable about all three axes at all speeds.

A.L.2. Part II Para. 41 (ii)

(ii) *Change of trim*

Undercarriage up	.. .. ..	Slightly nose up
Flaps up	.. ..	Slightly nose down
Bomb doors open	.. .. ..	Slightly nose up

There is considerable change of directional trim with changes of speed and power; therefore, for accurate flying, rudder trimming tab should be used to reduce heavy foot loads as necessary.

(iii) *Trimming tabs.*—The tabs on all three flying controls are powerful, particularly the elevator tabs.

(iv) *Controls.*—The rudder is heavier than on earlier Marks of Halifax aircraft, but elevator and aileron controls are much the same.

(v) *Flying at low airspeeds.*—The aircraft is pleasanter to fly at speeds below 140 m.p.h. I.A.S. with the flaps lowered 35°.

42. Stalling

(i) *Characteristics at the stall.*—There is a slight warning of the stall by a snatching of the ailerons occurring at about 5 m.p.h. above the stall. The stall is gentle and straight with no wing-dropping tendency; control is regained without difficulty on pushing the control column forward.

PART II—HANDLING INSTRUCTIONS

A.L.3
Part II
Paras.
42 (ii)
to 44

(ii) The stalling speeds in m.p.h. I.A.S. are:

	46,000 lb.	55,000 lb.
Flaps and undercarriage up	96 (106)	105 (115)
„ „ „ down ..	74 (87)	85 (98)
Flaps 30° down: undercarriage up ..	81	90

(iii) *Stalling in turns.*—There is considerable warning of the approach of a stall in turns to the left, but in turns to the right there is negligible warning, and at the stall the right wing drops sharply. Normal recovery action is effective in both cases, but must be applied immediately if the stall has occurred during a turn to the right to prevent the wing dropping further.
It should be noted that the stalling speed will be increased in turns, e.g., under an acceleration of 2g (the acceleration imposed in a sustained 60° bank turn) the stalling speed at 55,000 lb. will be approx. 150(160) m.p.h. I.A.S. and at 60,000 lb. it will be approx. 160(170) m.p.h. I.A.S.

43. **Diving**

(i) The aircraft becomes increasingly tail heavy as speed increases and should, therefore, be trimmed into and during the dive. There is a strong tendency for the left wing to drop and for the aircraft to yaw to the left, and this should be counteracted by use of the rudder trimming tab; on aircraft fitted with Type D tail turret and associated equipment this is most important. If, for any reason, the aircraft is not trimmed into the dive it will be necessary to adjust the elevator trimming tab, to avoid excessive forward force on the control column being necessary to prevent too rapid recovery.

44. **Approach and landing**

(i) Check with flight engineer:
Fuel contents of tanks in use.
Undercarriage up-locks disengaged and clips secured.
Flaps isolating cock open.
NOTE.—If there is a likelihood that the hydraulic system has been damaged, before opening the flaps isolating cock the engineer checks with the pilot that the flaps lever is in the " neutral " position and then turns the isolating cock half a turn. The pilot tells him immediately if the flaps are dropping—if so, the engineer can regulate their descent with the cock.

(ii) Reduce speed to 150 (160) m.p.h. I.A.S. and lower flaps 30°–40°; then reduce speed to 140 (150) m.p.h. I.A.S. and lower undercarriage.

(iii) *Check list for landing:*

Auto-pilot	OUT
Superchargers	M ratio
Air-intake	COLD AIR
Gills	As required
U—Undercarriage	DOWN
M—Mixture control (if fitted)	DOWN
P—Propeller	2,400 r.p.m.
F—Flaps	Fully down on final approach, or less in high wind

(iv) *Recommended speeds for the approach at 50,000 lb.:*

Engine assisted	110 (115) m.p.h. I.A.S.
Glide	120 (125) m.p.h. I.A.S.

PART II—HANDLING INSTRUCTIONS

45. **Beam approach** (static vent connected)

Stage	Indicated height ft.*	I.A.S. m.p.h.†	R.P.M.	Approx. boost	Actions	Change of trim
Preliminary approach	1,500	150	2,400	−3	Lower flaps 35° ..	Tail heavy
		140	2,400	−2½	Lower undercarriage on QDR at OMB	Slightly nose heavy
Outer marker beacon	700	135	2,400	−2	Maintain steady rate of descent	
			2,400	−1	Should give level flight	
Inner marker beacon	200	125	2,500		Lower flaps fully ..	No change
Overshoot	Up to 100 ft.	120	2,500	+6	Raise flaps to 40°, then raise undercarriage. Adjust boost and revs. at 1,000 feet	No change

* After adjusting altimeter for QFE and touch-down error as follows:
(i) With static vent connected, altimeter reads zero at touch-down with full flap.
(ii) With Mark VIII pressure head, altimeter reads +60 feet at touch-down with full flap, so subtract 2·2 millibars from QFE to give zero reading at touch-down.

†Subtract 5 m.p.h. for aircraft using Mark VIII pressure head.

PART II—HANDLING INSTRUCTIONS

46. Mislanding

The aircraft shows no change of trim when throttles are opened with flaps and undercarriage down unless the elevator trim has been wound fully back. Climb away at 100 (105) m.p.h. I.A.S., raise flaps to 40° down, and then raise undercarriage, then increase speed to 145–150 m.p.h. I.A.S.

NOTE.—If propellers are set to 2,400 r.p.m., set fully up immediately should it be necessary to use more than +6 lb./sq.in. boost.

47. After landing

(i) Before taxying, raise flaps; engineer opens gills.

(ii) When aircraft is taxied to dispersal the tailwheel should be left straight, thus eliminating unnecessary strain on the tailwheel centralising spring.

(iii) *Shutting down procedure:*

(*a*) Open up gradually and evenly, and run the engine for about 5 seconds at – 2 lb./sq.in. boost.

(*b*) Close the throttle slowly and evenly, taking about 5 seconds until speed is reduced to 800–1,000 r.p.m

(*c*) Run at this speed for a further two minutes.

(*d*) Operate the slow-running cut-outs by turning OFF the engine master cocks, and, *when the engine has stopped,* switch OFF the ignition.

(iv) Turn off all fuel tank cocks.
Switch off:

Pressure-head heater switch
Fuel contents gauges
DR compass and other equipment
Turn GROUND/FLIGHT switch to GROUND

(v) *Oil dilution.*—The correct dilution period for this aircraft is 4 minutes, and the dilution operation is to be carried out at an engine speed not exceeding 1,000 r.p.m.

AIR PUBLICATION 1719C & G—P.N.
Pilot's and Flight Engineer's Notes

PART III

OPERATING DATA

48. Engine data—Hercules VI or XVI

(i) *Fuel.*—100 octane only.

(ii) *Oil.—See* A.P. 1464/C37.

(iii) *Engine limitations.*—The *maximum* permissible r.p.m., boost and temperatures for the conditions of flight and the periods stated are:

		R.P.M.	Boost lb./sq.in.	Temp. °C. Cylinder	Temp. °C. Oil
TAKE-OFF TO 1,000 FT.	M	2,800	+8¼		
CLIMBING 1 HR. LIMIT	M S	2,400 2,500	+6	270	90
RICH CONTINUOUS	M, S	2,400	+6	270	80
*WEAK CONTINUOUS	M, S	2,400	+2	270	80
COMBAT 5 MINS. LIMIT	M, S	2,800	+8¼	280	100

*Weak mixture conditions are obtained on Hercules XVI engines at or below +2 lb./sq.in. boost.

OIL PRESSURE
NORMAL .. .. 80–90 lb./sq.in.
MINIMUM .. .. 70 lb./sq.in.

OIL TEMPERATURE FOR TAKE-OFF
RECOMMENDED .. .. 15°C.
MINIMUM .. .. 5°C.

MAXIMUM CYLINDER TEMPERATURE
TAKE-OFF .. .. 230°C.
STOPPING ENGINES .. .. 230°C.

PART III—OPERATING DATA

49. Flying limitations

(i) The aircraft is designed for manœuvres appropriate to a heavy bomber, and care must be taken to avoid imposing excessive wing loads in recovery from dives and turns at high speed. Spinning and aerobatics are not permitted. Violent use of the rudder should be avoided at high speeds.

(ii) *Maximum speeds in m.p.h. I.A.S.:*

Diving	320
Undercarriage DOWN	150
Flaps DOWN	150
Bomb doors OPEN	320

(iii) *Maximum weights:*

Take-off	63,000 lb.
Landing	55,000 lb.

(iv) *Bomb clearance angles:*

Diving 30°
Climbing 20°
Bank 10°—except for the 250 lb. " B " bomb, for which the maximum angle of bank is $2\frac{1}{2}$°.

50. Position error correction

(i) Aircraft with static side of pilot's A.S.I. connected to the Mark VIII pressure head.

From To	.. ..	120 140	140 160	160 180	180 210	210 240	m.p.h. I.A.S.
Add	..	12	10	8	6	4	m.p.h.

(ii) The position error correction for aircraft with static vent connected is + 1 m.p.h. at all speeds.

51. Maximum performance

(i) For maximum rate of climb:

(*a*) Climb at 130 (140) m.p.h. I.A.S. at full climbing power with gills fully open.

(*b*) Change to S gear when boost falls to + 3 lb./sq.in. and increase r.p.m. to 2,500.

(*c*) Close gills fully at 15,000 feet.

PART III—OPERATING DATA

(*d*) When boost falls to + 3 lb./sq.in. retract throttles to the position marked mid-way between RB and ECB.

(*e*) When boost falls to + 2 lb./sq.in. retract throttles to the ECB position and increase speed to 150 (160) m.p.h. I.A.S.

(ii) *All-out level.*—Use S ratio if maximum boost obtainable in M ratio drops below +5 lb./sq.in.

52. **Maximum range**

(i) Recommended operational climb:

(*a*) Climb at 150 (160) m.p.h. I.A.S. using maximum weak mixture cruising boost with gills closed. Select boost as follows: After take-off retract the four throttles together until boost falls just below +2 lb./sq.in., then progress throttles until + 2 lb./sq.in. exactly is regained. Check that weak mixture boost is not exceeded by comparison of the four exhaust glows.

(*b*) When boost falls to +1 lb./sq.in., change to S gear. Keep within cylinder temperature limitations by increasing speed where necessary, even to the extent of cruising in level flight for a short period. Do not progress throttles to the mid-position at any time.

(ii) *Cruising*

(*a*) Fly in M ratio at maximum obtainable boost not exceeding +2 lb./sq.in. obtaining the recommended airspeed by reducing r.p.m.

(*b*) The recommended speeds are:

Fully loaded	165 (175) m.p.h. I.A.S.
Lightly loaded	160 (170) ,, ,,

53. **Fuel capacity and consumptions**

(i) Capacity

2—No. 1 Inboard tanks	494 gallons
2—No. 2 Inner wing nose tanks ..	124 ,,
2—No. 3 Centre tanks	376 ,,
2—No. 4 Outboard tanks	322 ,,
2—No. 5 Outer engine tanks ..	244 ,,
2—No. 6 Outer engine tanks ..	246 ,,
	1,806 ,,

PART III—OPERATING DATA

(ii) Rich mixture consumption (approx.)—M ratio at 5,000 feet:

R.p.m.	Boost lb./sq.in.	Approx. total consumption galls./hr.
2,800	+8¼	640
2,400	+6	478

(iii) Weak mixture consumptions (approx.) in galls./hr.

Boost lb./sq.in.	M ratio at 5,000 ft. R.P.M.				S ratio at 15,000 fet. R.P.M.			
	2,400	2,200	2,000	1,800	2,400	2,200	2,000	1,800
+2	236	220	204	188	232	220	212	192
0	212	196	184	160	208	200	192	176
−2	188	176	164	148	188	180	172	160
−4	168	160	148	136	172	164	156	—

M gear:
For every 1,000 ft. above height quoted add 1 gall./hr.
" " " " below " " deduct 1 gall./hr.
S gear:
For every 1,000 ft. above height quoted add 2 galls./hr.
" " " " below " " deduct 2 galls./hr.

54. **A.S.I. conversion table**

M.p.h.	Knots	M.p.h.	Knots
74	64	130	113
80	70	135	117
81	71	140	122
90	78	145	126
95	82	150	130
100	87	155	134
105	91	160	139
110	96	165	143
115	100	170	148
120	104	200	174
125	109	320	278

AIR PUBLICATION 1719C & G—P.N.
Pilot's and Flight Engineer's Notes

A.L.2
Part IV
Paras.
55 & 56

PART IV—*EMERGENCIES*

55. **Engine failure on take-off**

(i) The aircraft can be kept straight on any three engines at take-off power at full load provided a safety speed of 145 (155) m.p.h. I.A.S. has been attained.

(ii) In the event of an outer engine failure below safety speed, control will be lost unless the opposite outer engine is immediately throttled back, at least partially. Feather the propeller of the failed engine, retrim, and reopen the throttle of the live outer engine.

(iii) After control has been retained as described above, it will be possible to climb with flaps in the take-off position and undercarriage up on three engines at take-off power at 130 (140) m.p.h. I.A.S. at light loads.
At heavy loads or if the engine failure has occurred at a low height immediately after take-off it will be necessary to land straight ahead using the two inner engines to control the rate of descent.

56. **Engine failure in flight**

(i) At full load, flaps and undercarriage up, at climbing power, control can be retained with rudder and aileron provided speed is not below 150 (160) m.p.h. I.A.S. If an outer engine fails below this speed it will be necessary to throttle back the opposite outer engine, at least partially, until rudder trim can be applied and the dead engine feathered. After trim has been applied, reopen the throttle of the live outer engine slowly: cases have occurred of propeller overspeeding due to a too rapid return of power after the propeller has fined its pitch.

(ii) *One engine failed.*—At 54,000 lb. use weak mixture cruising boost. Trim to fly without foot load at 140 (150) m.p.h. I.A.S. Height can normally be maintained to 18,000 feet.

(iii) *Two engines failed.*—Speed should not be allowed to fall below 135 (145) m.p.h. I.A.S.; lower speed will not reduce the rate of descent, and at this speed no undue strain is required to keep straight with two engines failed on one side. Fly in M gear and allow the aircraft to descend until height can be maintained. At light weights with symmetrical power available this may be possible in weak mixture up to about 12,000 feet, but at heavy loads or with two engines failed on one side, full climbing power may be necessary at about 5,000 feet.

(iv) The auto-pilot is of sufficient power to be used in all cases of engine failure except two engines failed on one side. The compressor, however, is driven by the port inner engine.
NOTE.—Increasing the r.p.m. on all engines will reveal the failed engine, but this test should not be carried out above cruising speed.

(v) *Landing on three engines.*—Lowering flaps to 35°, and undercarriage, may be carried out as in a normal circuit. Full flap should not be lowered nor rudder trim wound off until it is certain that the airfield can be reached comfortably on a straight approach. Final approach should be made at 120–125 m.p.h. I.A.S. using as little power as possible.

PART IV—EMERGENCIES

A.L.2
Part IV
Paras.
57 & 58

(vi) *Landing on two engines (asymmetric power).*—The circuit must be made with the two good engines on the inside of the turn, in which speed should not be allowed to fall below 140 (150) m.p.h. I.A.S. Operation of the undercarriage and flaps should be left as late as practicable. Aim at having the undercarriage locked down just before the final approach. Keep extra height if possible and approach in a glide at a speed of 125–130 m.p.h. I.A.S. Do not lower flaps nor wind off rudder trim until it is certain that the airfield can be reached comfortably in a glide. Some power may be required in the early stages of the approach.

(vii) *Propeller overspeeding.*—At low speeds the difficulties of control may be accentuated by an overspeeding windmilling propeller and will require special precautions. If overspeeding of any engine occurs:

(*a*) Close throttle of affected engine at once.

(*b*) Close throttle of corresponding engine on the other side in order to assist control.

NOTE.—Unless it is essential to retain height, the simplest immediate action is to close all four throttles.

(*c*) Normal corrective action should be taken (*see* A.P. 2095 Pilot's Notes General) but it should be noted that this may involve reducing speed below safety speed.

57. **Feathering and unfeathering**

(i) To feather a propeller:

(*a*) Hold the button (37) in only long enough to ensure that it stays in by itself.

(*b*) Close throttle immediately.

(*c*) Switch off only when engine has stopped and turn off master fuel cock.

(ii) To unfeather a propeller:

(*a*) Speed should not be greater than 160 (170) m.p.h. I.A.S.

(*b*) Set propeller control fully down and throttle slightly open.

(*c*) Turn on ignition and master fuel cock.

(*d*) Press and hold in feathering button until 1,000 to 1,300 r.p.m. is reached. If the propeller does not return to normal constant-speed operation, open throttle slightly.

58. **Undercarriage emergency operation**

(i) The mechanical up-locks must be released and the undercarriage control lever must be placed in the DOWN position.

NOTE.—The mechanism has been designed with a " weak link " so that in extreme emergency if the mechanical up-locks cannot be released or are inaccessible, the application of hydraulic power will break the lock and allow the undercarriage to lower.

(ii) If after releasing the mechanical up-locks and setting the undercarriage lever to DOWN the accumulator power will not lower the undercarriage, open the emergency cock on the front face of the front spar, leaving the undercarriage lever at DOWN. Either the engine-driven pump or handpump should then operate the undercarriage through the emergency pipe-lines.

(iii) In the event of complete failure of the hydraulic system the undercarriage will descend and lock down under its own weight assisted by the pull of the elastic cords fitted to each radius rod. Reduce speed if necessary.

PART IV—EMERGENCIES

59. **Bomb doors emergency operation**

Should the bomb doors fail to open when the pilot selects bomb doors OPEN:

(i) Check isolating cock is open.

(ii) Open the emergency cock on the front face of the front spar, and either the engine-driven pump or handpump should then operate the doors.

NOTE.—If the undercarriage is raised or the bomb doors closed after using either of the emergency circuits there may be insufficient fluid left in the system to lower the undercarriage a second time.

60. **Air-intake and landing lamps**

These hydraulic services can be operated by the hand-pump if the engine-driven pump fails. There are no separate emergency pipe-lines.

61. **Bomb jettisoning**

(i) The controls cannot be operated unless the bomb-door warning lights are showing.

(ii) Jettison bomb containers first, by pressing the button (44) under the flap directly below the left-hand warning light on the main instrument panel.

(iii) Jettison main load by pulling out the bomb-jettison handle (14) above the warning lights on the main instrument panel.

62. **Parachute exits**

(i) The hatch in the floor of the nose compartment.

(ii) The main entrance door of the port side of the rear fuselage.

(iii) The opening exposed by rotating the rear turret through 90°.

(iv) The paratroop cone (if fitted).

63. **Crash exits**

(i) The opening formed by raising or jettisoning a hinged transparent panel in the roof over the first pilot's seat.

(ii) The hatch in the fuselage roof aft of the front spar.

(iii) The hatch in the fuselage roof aft of the rear spar. A folding escape ladder is fitted below this exit.

PART IV—EMERGENCIES

EMERGENCY EQUIPMENT

64. **Fire-extinguishers.**—A semi-automatic fire-extinguisher system is installed, and is operated by gravity and impact switches. The system may be operated manually by manipulating the pilot's pushbutton switches (65) mounted on the port side of the cockpit. There are four such buttons, one for each engine bay. Hand-operated fire-extinguishers are stowed in the following positions:

Type No. 3 *fire-extinguishers:*

On the fuselage roof above the navigator's position.

In the roof of the flight engineer's position, port side of the astral dome.

Type No. 5 *fire-extinguishers:*

On the forward face of the pilot's bulkhead to the pilot's right hand.

Starboard side of the fuselage forward of the main electrical panel.

Above the starboard rest seat.

Starboard side of the fuselage forward of the instrument panel at the flare launching station.

Starboard side of the fuselage just forward of the rear gun turret.

65. **Dinghy.**—A dinghy stowed on the port side in the centre plane is released by

(*a*) A manual release on the port side of the rear roof escape hatch. Give handle a half turn counter clockwise and pull.

(*b*) Immersion switches under the nose of the fuselage.

A signal pistol and cartridges, emergency rations, sea markers, first-aid outfit and paddles are provided in a valise attached to the dinghy by a cord.

(*c*) Ditching should be carried out with flaps 35° down.

AIR PUBLICATION 1719C & G—P.N.
Pilot's and Flight Engineer's Notes

PART V

ADDITIONAL DATA FOR FLIGHT ENGINEER

66. **Damage by enemy action**

The flight engineer must carry out the drills given in A.P. 2764 Flight Engineer's Notes General, and in addition must check:

(*a*) All hydraulic accumulator pressures.

(*b*) All oxygen equipment.

67. **Landing away from base**

The flight engineer proceeds as detailed in A.P. 2764 Flight Engineer's Notes General, but should the landing have been made due to fuel shortage only:

(*a*) Check GROUND/FLIGHT switch is at GROUND and pitot head cover is on.

(*b*) Instruct duty crew to fill an even amount of fuel into tanks 1, 2, 3 and 4 (provided take-off weight is below 55,000 lb., *see* para. 33 (*g*) sufficient for the flight with a safety margin depending on the weather conditions. Supervise filling procedure.

(*c*) Instruct duty crew to check oil. Check grade of oil and supervise filling procedure.

(*d*) Check brake pressure and have system recharged if necessary.

(*e*) Arrange that 24-volt ground battery is available for starting.

(*f*) Carry out pre-flight check.

(*g*) Ground test engines if necessary. Give instructions, before starting, of drill used.

NOTE.—The carburettor must not be primed unless it is known that the float chambers are empty, in which case not more than three strokes of the pump must be given.

PART V—DATA FOR FLIGHT ENGINEER

68. Hydraulic system

Tank.—The hydraulic fluid is stored in a tank in the starboard engine nacelle. The tank is provided with a filling filter, the bottom of which serves as a filling level for the fluid.

Handpump.—A double-acting pump is attached to the front spar on the port side in the fuselage. The pump draws fluid from the tank return pipe and will operate the entire system in the same manner as the engine pump.

Distributors.—The distributors for the undercarriage, flaps and bomb doors are remotely controlled through linkage members by a group of hand levers situated to the right of the pilot's seat. The levers have a hidden three-position gate; the central position being neutral. The flaps may be stopped in any desired intermediate position by returning the lever to this neutral position. The landing lamp distributor lever is spring-loaded to remain in the neutral position and has to be held in the operating position while the operation is in progress. The air-intake distributor lever has two positions only.

Isolating valves.—A manually-operated isolating valve is fitted between the top of the flap jacks and accumulator and one in the accumulator line leading to the underside of all the bomb-door jacks. These isolating valves, when closed, prevent the complete lowering of the flaps and bomb doors by accumulator pressure, in the event of damage to the pipes. When the pilot's levers are selected to " doors open" and " flaps down " with their respective isolating valves closed, however, this has the same effect as breaking the pipe-lines on the opposite side of the jack to its accumulator and allows the weight of the flaps or bomb doors to react against the jacks, thereby forcing fluid from the jacks to return through the distributor back to the tank. The amount of opening and lowering varies according to conditions. If the pilot's levers are selected when on the ground as stated above, the flaps will tend to lower themselves fully and the bomb doors partly open. In flight, however, the flaps will tend to fall only slightly, and the bomb doors will open a varying

PART V—DATA FOR FLIGHT ENGINEER

amount according to conditions. It is important, therefore, that the pilot's levers be returned to neutral after a particular circuit has been operated.

Mechanical up-locks.—The controls for the mechanical up-locks are above the rest seats one on either side of the fuselage. The red lights on the pilot's undercarriage indicator remain on until the up-locks are engaged by pushing in the control handles.

Accumulators.—The accumulators, with the exception of the power accumulator, are not fitted with separator pistons, as the circuits are sealed by the jack pistons, thus preventing leakage of air into the system. The locations of the accumulators are as follows:

(i) Undercarriage accumulators are attached to the front spar aft of the inner engines.

(ii) Bomb-door accumulator is aft of the front spar, attached to the port side of the fuselage.

(iii) Flaps accumulator is aft of the rear spar, on the starboard side of the fuselage.

(iv) Power accumulator is mounted on the front spar in the starboard inner nacelle.

(v) Tailwheel accumulator (fitted on later aircraft) is aft of the rear fuselage bay bulkhead.

Gauge relay.—The gauge relay is fitted in the pipe-line to the pressure gauge of the main hydraulic circuit and is designed to isolate the gauge in the event of a fault in the gauge or the pipe-line. The gauge is adjacent to the handpump.

Engine pump.—This is of the multi-cylinder high-speed radial type. The pump runs continuously, the strain on the pump being relieved between operational demands by an hydraulic cut-out valve. The pump is mounted on the gear box of the starboard inner engine.

Cut-put valve.—An automatic cut-out valve is fitted in the pump delivery line, and is mounted on the rear face of the starboard inner engine diaphragm. When the

PART V—DATA FOR FLIGHT ENGINEER

pump has charged the power accumulator to 2,400–2,500 lb./sq.in. the valve isolates the system from the pump and by-passes the fluid back to the tank. The valve continues to by-pass fluid until the accumulator pressure has dropped to 2,000 lb./sq.in.

Pressure limiting valve.—This valve is fitted to limit the pressure obtainable by the pump to 2,800 lb./sq. in.

Safety valves.—A safety valve, fitted in each undercarriage accumulator line, limits the pressure which can be applied to the accumulator in the event of damage to the hydraulic locks or jacks, to 600 lb./sq. in.

Flexible pipes.—The flexible pipes are steel-cored high-pressure hoses. They are suitable only for use with the specified hydraulic fluids and should not be brought into contact with other fluids, particularly mineral oils such as paraffin and petrol.

69. **Pneumatic system**

Air services.—The air supply is obtained from a Heywood compressor driven by the port inner engine. The air passes through a pressure regulator mounted behind the compressor, then through an oil trap and along the false spar to the air bottle in the fuselage. The air bottle is situated on the port side of the fuselage aft of the engineer's armour plate bulkhead. When the air bottle is charged to the normal working pressure of 300 lb./sq.in. the regulator cuts off the supply to the air bottle.

Brake system.—The air from the bottle passes through an air filter to the differential control unit and thence to the brakes. For ground filling purposes the air-charging valve is situated in the rear of the port inner engine nacelle and is accessible through the undercarriage doors.

Wheel brake unit.—The wheel brake unit is a Dunlop component and is described in A.P. 2337.

70. **Engine-driven accessories**

The engine-speed indicator generator and the constant-speed unit are part of the basic power plant. Accessories are fitted as follows.

(i) *Starboard outer:*

(*a*) A 500-watt 24-volt A.C. generator to supply the special A.R.I.s.

PART V—DATA FOR FLIGHT ENGINEER

(ii) *Starboard inner:*

(*a*) A 1,500-watt 24-volt D.C. generator. (Alternatively a 1,000-watt D.C. generator with a 24-volt alternator driven in tandem where special A.R.I.s are fitted.)

(*b*) A Pesco vacuum pump to supply the Mark XIV bomb-sight, or the instrument flying panel in emergency.

(*c*) A Lockheed pump for the hydraulic system.

(iii) *Port inner:*

(*a*) A 1,500-watt 24-volt D.C. generator.

(*b*) A Pesco vacuum pump to supply the instrument flying panel.

(*c*) A Heywood compressor for the pneumatic system.

(*d*) An R.A.E. compressor for the supply to the automatic pilot.

(iv) *Port outer:*

(*a*) A 1,500-watt 24-volt D.C. generator.

NOTE.—All the D.C. generators are connected in parallel and supply the lighting and general services of the aircraft.

71. **Main pressures**

	lb./sq.in.
R.A.E. compressor to auto-pilot	60
Engine oil pressure (minimum)	70
,, ,, ,, (normal)	80–90
Brake pressure (minimum)	90
Inflation pressure undercarriage accumulator	250
Brake pressure (supply)	300
Inflation pressure bomb doors accumulator (OPEN)	700
,, ,, flaps accumulator (flaps DOWN)	400
,, ,, power accumulator ..	1,850
Lockheed cut-out operation (in)	2,000
,, ,, ,, (out) ..	2,400–2,500
Pressure limiting valve setting	2,800

AIR PUBLICATION 1719C & G—P.N.
Pilot's and Flight Engineer's Notes

PART VI

LOCATION OF CONTROLS AND ILLUSTRATIONS

AIR PUBLICATION 1719C & G—P.N.
Pilot's and Flight Engineer's Notes

PART VI

ILLUSTRATIONS AND LOCATION OF CONTROLS NOT ILLUSTRATED

Page 44

LOCATION OF CONTROLS

E. denotes controls operated by flight engineer.

Aircraft controls

Control	Location
Aileron locking gear	Bag on starboard side aft of engineer's armour plate door
Rudder and elevator locking gear	Bag on port side of rear fuselage forward of rear turret

Fuel system

Control	Location
Master engine cocks	Bulkhead aft of pilot
E. Tank selector cocks (Mark III)	Forward end of rest seats
E. Wing crossfeed cocks	Forward end of rest seats (Mark III)
E. Centre crossfeed cock ..	Aft face of rear spar
E. Long-range fuel distribution cocks	Under step aft of front spar
Fuel pressure warning lights ..	Engineer's panel Duplicated at rest station (Mark III)
Priming pump	Rear of each undercarriage fairing (Mark III) Engineer's station (Mark VII)

Hydraulic system

Control	Location
E. Undercarriage mechanical up-lock controls	Above rest seats
E. Flaps isolating cock	Central on aft face of rear spar
E. Bomb doors isolating cocks ..	Under bomb doors accumulator port side, aft of front spar
E. Bomb doors selective closing cock	Aft of pilot's bulkhead
E. Hydraulic handpump	Port side, front spar
E. Undercarriage emergency cock	Front face, front spar
E. Bomb doors emergency cock ..	,, ,, ,, ,,

Electrical controls

Control	Location
E. GROUND/FLIGHT switch ..	Starboard side of fuselage opposite engineer's station
Ground battery connection ..	Starboard side under leading edge
Electrical control panel	Right-hand side of W/T operator's station
Main generator fuses	Right-hand side of W/T operator's station
Generator warning lights ..	Right-hand side of W/T operator's station
Voltmeter and ammeter ..	Right-hand side of W/T operator's station

Operational controls and equipment

E. Heating controls	On ducts aft of front spar
E. Oxygen main valve	At foot of engineer's panel
E. Signal pistol	Fuselage roof aft of pilot's bulkhead
Signal pistol cartridges stowage	Rear face of bulkhead
Flying rations	Aft of engineer's bulkhead
Vacuum flasks	Aft of engineer's bulkhead
Drinking water	Aft of engineer's bulkhead
Aldis lamp	Bomb aimer's station

Emergency equipment

Fire-extinguishers	See Part IV
Dinghy manual release	Adjacent to rear roof escape hatch
Dinghy emergency pack	Port side, aft of rear spar
Dinghy emergency radio ..	Port side, above rest seats
Crash axes (three)	Starboard side, opposite flight engineer's station Starboard side, aft of rear spar Rear turret door
First-aid outfits (three)	Two on port side just forward of entrance door One on starboard side, bomb aimer's station
Emergency rations	Rear fuselage at the step up to floor over bomb-bay
Emergency oxygen	All crew stations
Incendiary bombs (aircraft) ..	Cover plate to engineer's panel

Key to *Fig. 1*

INSTRUMENT PANEL

1. Camera warning light
2. Oxygen flow meter
3. D.R. compass repeater
4. Flaps indicator
5. A.R.I. destruction switches (under cover)
6. Beam approach indicator
7. Bomb steering indicator
8. Instrument flying panel
9. Boost gauges
10. Suction gauge
11. Landing lamp switch
12. D.F. indicator
13. Engine-speed indicators
14. Bomb jettison handle
15. Bomb doors warning lights
16. Suction changeover cock
17. Troop signalling lamps and switch
18. Throttle friction adjusting lever
19. Brakes and supply pressure gauge
20. Landing lamps dipping lever
21. Bomb-firing switch
22. Starboard ignition switches
23. Undercarriage indicator
24. Port ignition switches
25. Horn warning light
27. Air temperature gauge
28. Auto-pilot pressure and trim gauge

FIG 1
INSTRUMENT PANEL
1
2
3
4
5
6
7
8
9
10
11
12
13
14
15
16
17
18
19
20
21
22
23
24
25
27
28
DE HAVILLAND
CONSTANT SPEED AIRSCREWS
HIGH REVS
LOW REVS
44200

Key to *Fig. 2*

GENERAL VIEW OF CABIN

29. Undercarriage lever
30. Flaps lever
31. Bomb doors lever
32. Elevator trimming tab control
33. Rudder pedals adjusting starwheel
34. Compass
35. Wheel brakes lever
36. "Press to transmit" switch
37. Feathering buttons
38. Recognition lights switchbox
39. Navigation lights switch
40. Pressure-head heater switch
41. Headlamp switch
42. Formation lights switch
43. Throttle levers
44. Bomb containers jettison switch
45. Oxygen regulator
46. Propeller speed control levers
47. Glider release control
48. Mixture lever
49. Superchargers control
50. Windscreen de-icing pump
51. Elevator trimming tab indicator

Key to *Fig. 3*

FLIGHT ENGINEER'S PANEL

52. Cowling gills, motor controls and indicators
53. Gills position indicators
54. Booster-coil buttons
55. Engine starting buttons
56. Fuel contents gauges
57. Fuel contents gauges switch
58. Oil temperature gauges
59. Cylinder temperature gauges
60. Oil pressure gauges
61. Fuel pressure warning lights
62. Oil dilution switches
63. Fuel transfer pumps ground testing buttons
64. Immersed pump switches

FIG. 3 FLIGHT ENGINEER'S PANEL FIG. 3

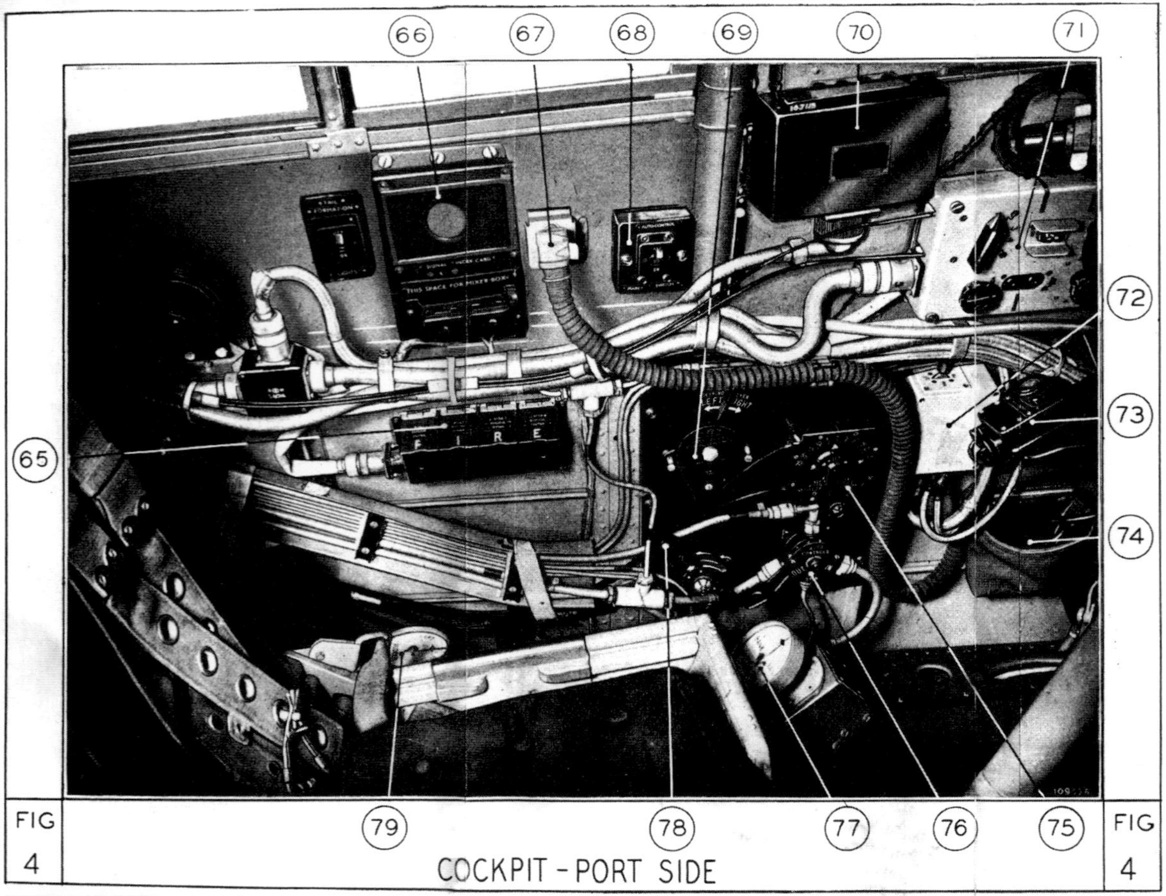

FIG 4

COCKPIT - PORT SIDE

Key to *Fig. 4*

COCKPIT—PORT SIDE

65. Fire-extinguisher pushbuttons
66. Signal index card holder
67. Pilot's oxygen connection
68. Auto-pilot main switch
69. Auto-pilot steering lever
70. T.R.1196 control box
71. Beam approach control unit
72. Beam approach mixer box
73. Pilot's call light
74. Heater socket stowage
75. Auto-pilot attitude control
76. Auto-pilot control cock
77. Aileron trimming tab control
78. Auto-pilot clutch lever
79. Rudder trimmer tabs control

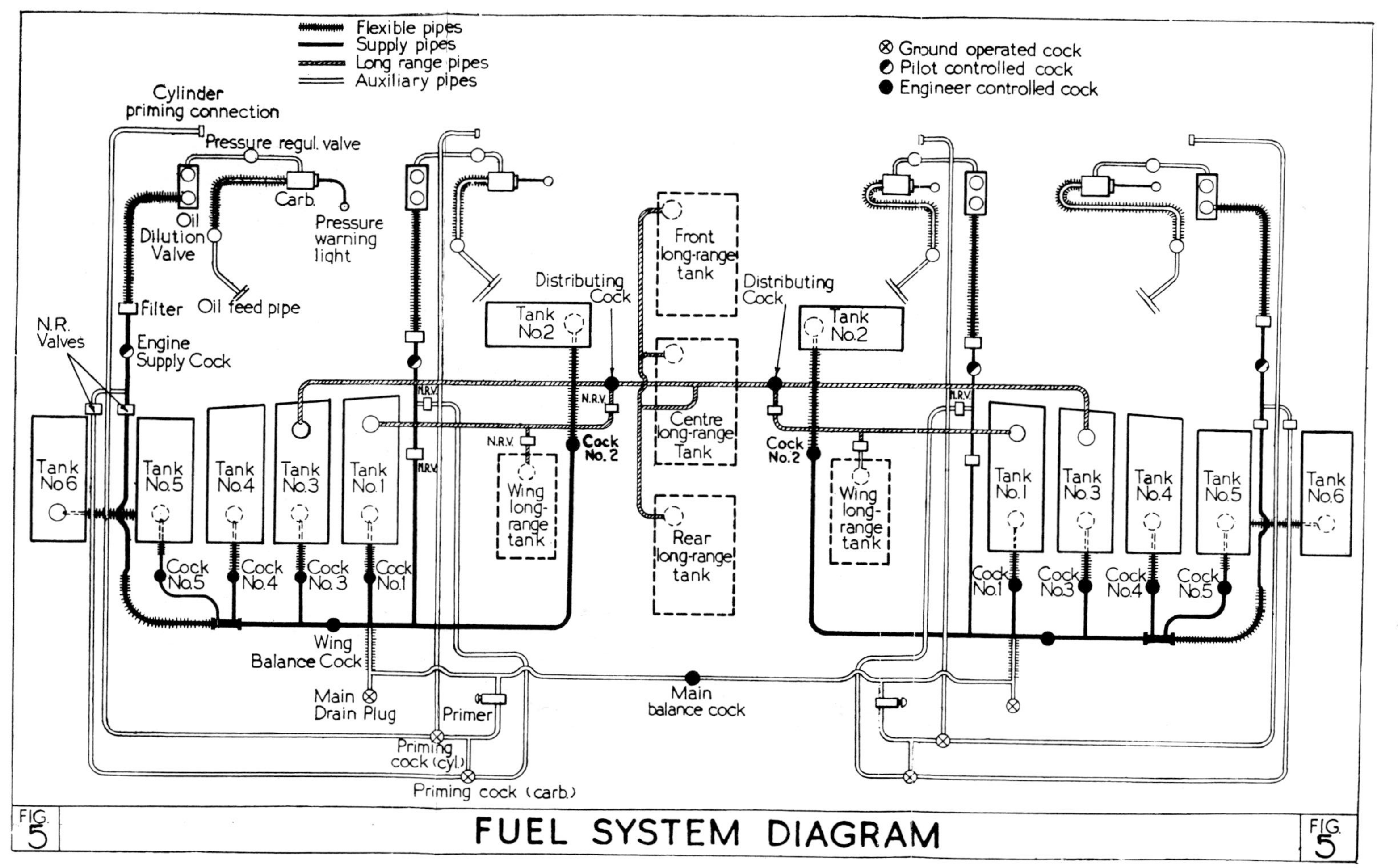
Flexible pipes
Supply pipes
Long range pipes
Auxiliary pipes
Ground operated cock
Pilot controlled cock
Engineer controlled cock
Cylinder priming connection
Pressure regul. valve
Carb.
Pressure warning light
Oil Dilution Valve
Oil feed pipe
Filter
N.R. Valves
Engine Supply Cock
Tank No.6
Tank No.5
Tank No.4
Tank No.3
Tank No.1
Tank No.2
Cock No.5
Cock No.4
Cock No.3
Cock No.1
Cock No.2
N.R.V.
Distributing Cock
Wing long-range tank
Front long-range tank
Centre long-range Tank
Rear long-range tank
Wing Balance Cock
Main Drain Plug
Primer
Main balance cock
Priming cock (cyl.)
Priming cock (carb.)
FIG. 5
FUEL SYSTEM DIAGRAM

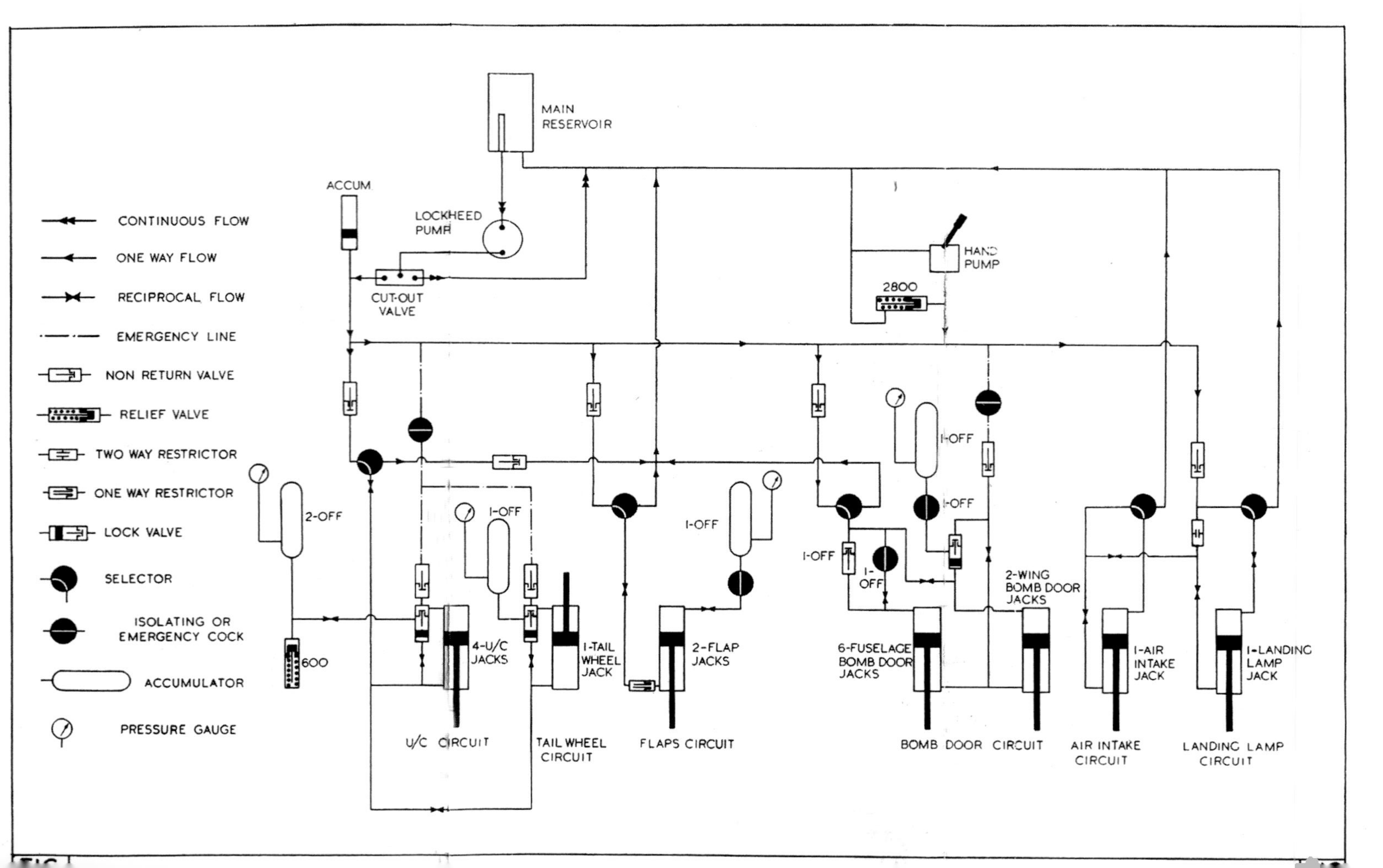
MAIN RESERVOIR
ACCUM
LOCKHEED PUMP
CUT-OUT VALVE
HAND PUMP
2800
CONTINUOUS FLOW
ONE WAY FLOW
RECIPROCAL FLOW
EMERGENCY LINE
NON RETURN VALVE
RELIEF VALVE
TWO WAY RESTRICTOR
ONE WAY RESTRICTOR
LOCK VALVE
SELECTOR
ISOLATING OR EMERGENCY COCK
ACCUMULATOR
PRESSURE GAUGE
2-OFF
600
1-OFF
4-U/C JACKS
1-TAIL WHEEL JACK
1-OFF
2-FLAP JACKS
1-OFF
1-OFF
1-OFF
1-OFF
1-OFF
6-FUSELAGE BOMB DOOR JACKS
2-WING BOMB DOOR JACKS
1-AIR INTAKE JACK
1-LANDING LAMP JACK
U/C CIRCUIT
TAIL WHEEL CIRCUIT
FLAPS CIRCUIT
BOMB DOOR CIRCUIT
AIR INTAKE CIRCUIT
LANDING LAMP CIRCUIT

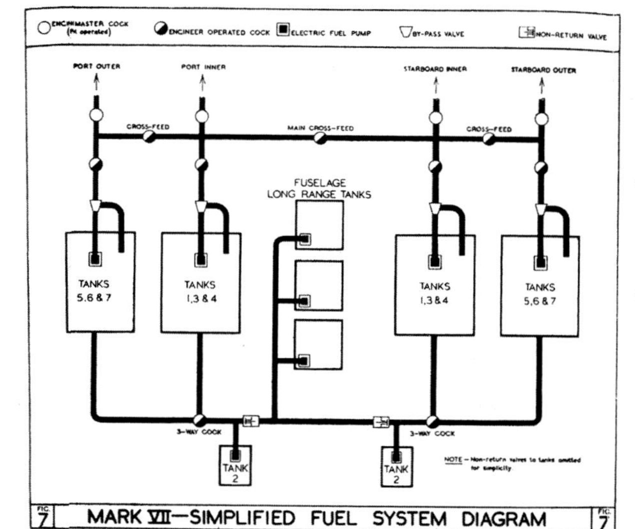

FIG 7 MARK VII—SIMPLIFIED FUEL SYSTEM DIAGRAM FIG 7

1. *Fuel tanks.*—Fuel is carried in fourteen self-sealing tanks in the wings. The seven tanks in each wing are so connected that they may be considered as two groups of three tanks, each group feeding one engine, and one replenishment tank from which fuel may be transferred into either group in that wing.
The capacities in each wing are:

Group	Tank	Capacity	Group total
Inboard group	Tank 1	62 gal.	
	Tank 3	247 gal.	497 gal.
	Tank 4	188 gal.	
Replenishment tank	Tank 2	150 gal.	150 gal.
Outboard group	Tank 5	161 gal.	
	Tank 6	122 gal.	448 gal.
	Tank 7	165 gal.	
		Total	1,095 gal.

Provision is made for fitting three self-sealing 230 gallon long-range tanks in the fuselage bomb-bay. A nitrogen fire protection system for all tanks is installed; nitrogen being fed into the tanks automatically as fuel is used, so that no inflammable petrol-air mixture is present in the tanks. The control valve is on the port side of the fuselage at the rest station and must be fully opened before any petrol is used

2. *Master engine cocks.*—The supply to each engine can be shut off by the appropriate master engine cock control lever mounted on the bulkhead aft of the pilot. Moving the levers to the down position closes the cocks and also operates the slow-running cut-outs.

3. *Fuel cocks.*—The layout of tanks and cocks is symmetrical about the centre line of the aircraft and the system in each wing should be operated independently with the centre crossfeed cock on the aft face of the rear spar kept closed. Each group of tanks is controlled by one ON-OFF cock; a crossfeed cock is fitted in the gallery pipe between the inner and outer groups and when this crossfeed cock is closed the inboard group feeds the inner engine and the outboard group the outer engine. Controls (103 & 104) for all these cocks are provided at the flight engineer's station below the instrument panel. Two three-way cocks are provided at the rear of the front spar to select the group of tanks into which fuel is to be transferred from the replenishment tanks. These cocks are also used when transferring fuel from the bomb bay tanks. The arrangement of the non-return valves is such that when both cocks are open (either group can be selected) the fuel will be transferred to both wings simultaneously from the long-range tanks, whereas fuel can only be transferred from the replenishment tanks into a group in the same wing, and any balancing of fuel from wing to wing must be done through the centre crossfeed cock.

4. *Electric pumps*

(i) Each group of tanks is provided with an electric booster pump. These pumps are fitted in a well compartment partitioned off at the rear of tanks 3 and 5 and into which the other tanks of the group drain. A by-pass valve on each pump limits the pressure in the feed pipe to 10 lb./sq.in. and returns the excess fuel being pumped, back to the tank. The by-pass valve also allows fuel to be drawn direct to the engine-driven pump if the electric pump is not working.

(ii) Electric transfer pumps are provided in each replenishment tank and each long-range tank and are used to transfer fuel to a selected group, or groups of tanks.

(iii) Switches to all pumps are mounted at the bottom of the flight engineer's panel together with ground testing equipment.

5. *Fuel contents gauges.*—The fuel contents of all wing tanks are registered on gauges, one for each tank, on the flight engineer's panel; but as described in para. 1, tanks 1, 3 and 4 can only be used as a unit and tanks 5, 6 and 7 as another, since no provision is made for isolating individual tanks. A circuit switch is provided adjacent to the gauges.
The long-range tanks are fitted with direct reading contents gauges visible through the bomb hatches in the fuselage floor.

6. *Fuel pressure warning lights:*

(i) Four fuel pressure warning lights are mounted on the flight engineer's panel and indicate low fuel pressure at the carburettors.

(ii) A fuel pressure warning light (95) for each replenishment tank and one (83) for the three long-range tanks are mounted on the lower portion of the flight engineer's panel. These lights warn when a tank is empty and the relevant pump should be switched off.

7. *Priming.*—A handpump for induction priming is mounted on a bracket on the port side of the fuselage at the flight engineer's station, and serves all engines. The supply of fuel is taken from tank 1 in the port wing. Provision is made for the use of high volatility fuel to facilitate starting in cold weather; the cock and connection is adjacent to the priming pump. There is no handpump for priming the carburettors and this must be done by operating the electric pump in the relevant group of tanks. The pump should not be left running while the engine is stationary.

8. *Management of fuel system.*—For normal operation one group of tanks should feed one engine, i.e. all tanks ON, and all crossfeed cocks OFF. The outboard group of tanks contains approx. 50 gallons less fuel than the inboard group and should be replenished from tank 2 as required by setting the transfer cock to the outboard group and switching on the pump in No. 2 tank. Take-offs should be made with the tank group pumps on and these pumps should also be used to augment fuel pressure at altitude (approx. 9,000 feet).

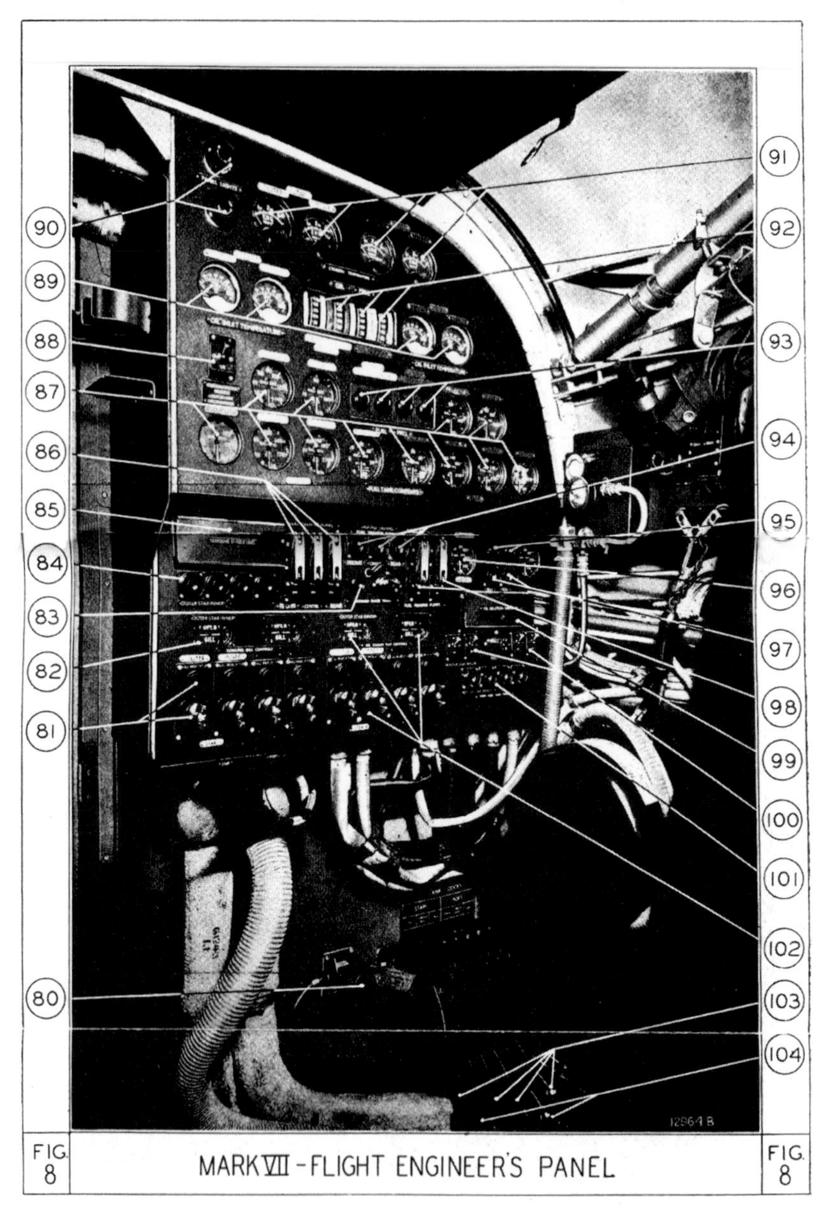

MARK VII - FLIGHT ENGINEER'S PANEL

Key to *Fig. 8* — FLIGHT ENGINEER'S PANEL

80. Main oxygen valve
81. Cowling gills motor controls and indicators
82. Gills position indicators
83. Long-range fuel pressure warning light
84. Booster-coil buttons
85. Engine starting buttons
86. Long-range fuel pump switches
87. Fuel contents gauges
88. Fuel contents gauges switch
89. Oil temperature gauges
90. Panel light switches
91. Cylinder temperature gauges
92. Oil pressure gauges
93. Fuel pressure warning lights
94. Long-range fuel pumps ground test buttons
95. Replenishment fuel pressure warning lights
96. Replenishment tanks fuel contents gauges
97. Warning light test buttons
98. Replenishment tanks pump switches
99. Oil dilution switches
100. Tank group pump switches
101. Tank group pump ground test buttons
102. Oil cooler flaps motor controls, indicators and position indicators (inoperative on this Mark of Halifax)
103. Tank group fuel cock control levers
104. Crossfeed cock control levers

AVRO LANCASTER

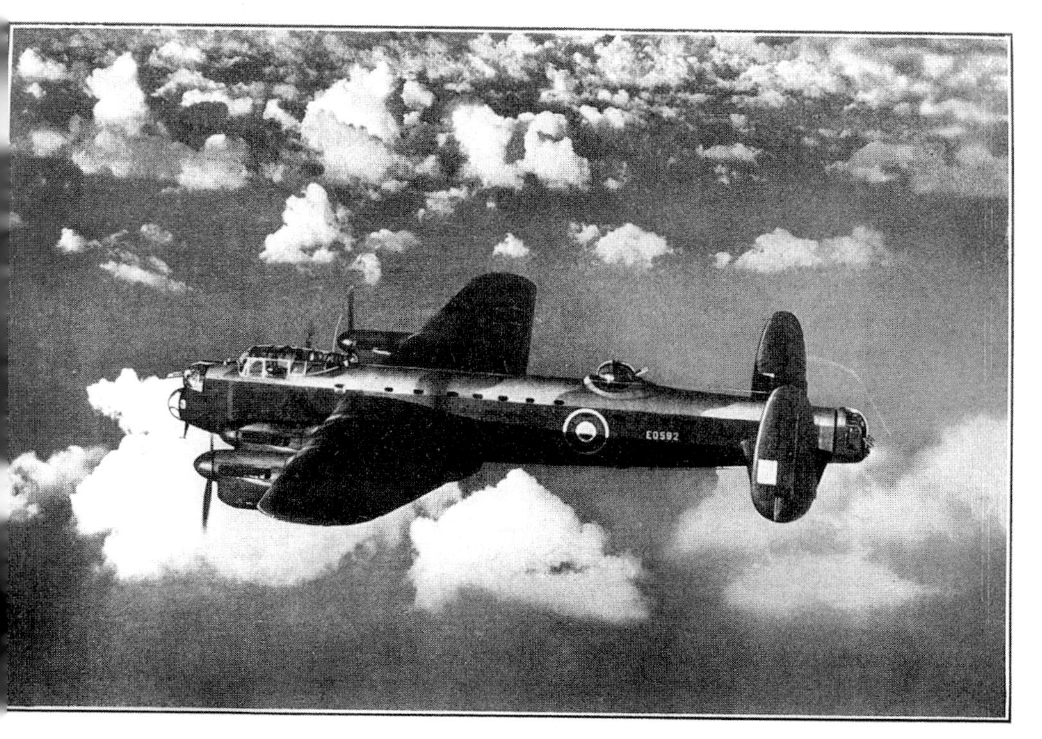

Photo by Charles E. Brown.

NOTES TO USERS

THESE Notes are complementary to A.P. 2095 Pilot's Notes General and assume a thorough knowledge of its contents. All pilots should be in possession of a copy of A.P. 2095 (see A.M.O. A718/48).

Additional copies may be obtained by the station publications officer by application on Form 294A, in duplicate, to Command headquarters for onward transmission to A.P.F.S. (see A.P.113.) The number of this publication must be quoted in full—A.P. 2062, A, C, F, H, K, L & M.—P.N.

Comments and suggestions should be forwarded through the usual channels to the Air Ministry (T.F.2.).

AIR MINISTRY A.P. 2062, A, C, F, H, K, L & M—P.N.
March, 1949 *Pilot's Notes*
(*4th Edition*)
(Reprinted—October, 1949)

LANCASTER Mks. 1, 3, 7 and 10

PILOT'S NOTES

LIST OF CONTENTS

ENGINE CONTROLS

OTHER EQUIPMENT

PART II—HANDLING

PART III—LIMITATIONS

PART IV—EMERGENCIES

PART V—ILLUSTRATIONS AND LOCATION OF CONTROLS

LANCASTER
(ALL MARKS)
PILOT'S CHECK LIST
(Excluding Checks of Operational Equipment)

	ITEM	CHECK
1.	Weight and balance.	Within permissible limits.
2.	Authorisation book.	Sign.
3.	Form 700.	Sign.

External checks.

N.B.—Start at the entrance door and work clockwise around aircraft.

	ITEM	CHECK
4.	Dinghy external release.	Secure.
5.	Tail wheel.	Extension of oleo leg. Tyre for cuts and creep. Valve free.
6.	Starboard tailplane.	Condition of upper and under surfaces. Leading edge.
7.	Starboard fin.	Condition. Leading edge.
8.	Starboard rudder.	Condition. Trimmer. External control lock removed.
9.	Starboard elevator.	Condition. Trimmer. External control lock removed.
10.	Rear turret.	Locked fore and aft.
11.	Rear lights.	Condition.
12.	External aerials.	Condition.
13.	Port elevator.	Condition. Trimmer. External control lock removed.
14.	Port rudder.	Condition. Trimmer. External control lock removed.
15.	Port fin.	Condition. Leading edge.
16.	Port tailplane.	Condition of upper and under surfaces. Leading edge.
17.	Static vent.	Plug removed.
18.	Port mainplane.	Condition of upper surface. Fuel tank filler covers secure.
19.	Port flaps.	Position. Condition.
20.	Port aileron.	Condition. Trimmer. External control lock removed.
21.	Port navigation light.	Condition.
22.	Port identification lights.	Condition.
23.	Port mainplane.	Condition of leading edge. Condition of undersurface.
24.	Landing lamps.	Condition. Position.
25.	No. 1 engine.	Security of oil tank and coolant filler covers. Security of cowlings. Condition of propeller and spinner. Oil and coolant leaks.

	ITEM	CHECK
26.	No. 2 engine.	Security of oil tank and coolant filler covers. Security of cowlings. Condition of propeller and spinner. Oil and coolant leaks.
27.	Port under-carriage.	External lock removed. Microswitches clean and free. Extension of oleo legs. Brake leads secure. Towing shackle secure. Locking ring for creep and wedges in position. Valve free. Tyre for cuts and creep. Chock in position.
28.	Port centre section.	Leading edge panel secure. All screws flush fitting.
29.	Port fuel jettison pipe cover plate.	Pin secure.
30.	Trailing aerial fair lead.	Condition.
31.	Bomb doors.	Condition.
32.	Pressure head.	Cover removed.
33.	Nose.	Condition.
34.	External fire-extin-guishers.	In position.
35.	Starboard centre section.	Leading edge panel secure. All screws flush fitting.
36.	Starboard fuel jettison pipe cover plate.	Pin secure.
37.	Starboard under-carriage.	External lock removed. Microswitches clean and free. Extension of oleo legs. Brake leads secure. Towing shackle secure. Locking ring for creep and wedges in position. Valve free. Tyre for cuts and creep. Chock in posi-tion.
38.	No. 3 engine.	Security of oil tank and filler covers. Security of cowlings. Condition of propeller and spinner. Oil and coolant leaks.
39.	No. 4 engine.	Security of oil tank and filler covers. Security of cowlings. Condition of propeller and spinner. Oil and coolant leaks.
40.	Starboard mainplane.	Condition of leading edge. Condition of undersurface.
41.	Starboard navigation light.	Condition.
42.	Starboard identifica-tion lights.	Condition.

	ITEM	CHECK
43.	Starboard aileron.	Condition. Trimmer. External control lock removed.
44.	Starboard flap.	Position. Condition.
45.	Starboard mainplane.	Condition of upper surface. Fuel tank filler covers secure.
46.	Static vent.	Plug removed.
47.	Dispersal area.	All clear around aircraft.

Internal checks.

N.B.—Start at the rear of the aircraft and work forward.

	ITEM	CHECK
48.	Rear turret.	Doors closed. Position of "dead man's" handle.
49.	Fire-extinguisher.	In position.
50.	Elsan.	Secure.
51.	Dipsticks.	In position.
52.	First-aid kit.	In position.
53.	Flying control rods and trimmer cables.	Free from obstructions.
54.	D.R. compass master unit.	Clear of magnetic interferences.
55.	Loose equipment.	All secured.
56.	Crash axe.	In position.
57.	Fire-extinguisher.	In position.
58.	Rear escape hatch (if fitted).	Operation. Secure.
59.	Mid upper turret.	Condition.
60.	Crash axe.	In position.
61.	Emergency pack.	Stowed.
62.	Dinghy radio.	Stowed.
63.	Main oxygen supply.	On.
64.	Nitrogen main cock (if fitted).	On if required.
65.	Mid escape hatch.	Operation. Secure.
66.	Hydraulic accumulator.	Pressure 220 lb./sq. in.
67.	Emergency air system.	Pressure (if gauge fitted) 1,200 lb./sq. in. approx.
68.	Ground/flight switch.	Flight.
69.	Electrical panel.	All circuit breaker switches in (Mk. 10 only). Battery state. Generator warning lights on. Generator switches on. Earth lights.
70.	Batteries.	Leads secure.
71.	Intercom.	Switch on.
72.	Cross-feed cock.	Freedom of movement. Turn off.
73.	Fire-extinguisher.	In position.
74.	Fuel gauges.	Switch on (Mks. 1, 3 and 7). Contents.
75.	Booster pumps.	Test operation by ammeter and switch off.

	ITEM	CHECK
76.	Pressure-head heater.	Test operation by ammeter. Switch off.
77.	Tank selector cocks.	As required.
78.	Fuel flowmeters (if fitted).	Set to zero.
79.	Emergency air selector knob.	Off.
80.	U/c warning horn and light.	Operate test pushbutton.
81.	Fire-extinguisher.	In position.
82.	Forward upper escape hatch.	Operation. Secure.
83.	Main parachute escape hatch.	Operation. Secure.
84.	Front turret.	Condition. Doors closed.
85.	Fire-extinguisher (in nose).	In position.
86.	U/c selector lever.	Down. Safety bolt engaged.
87.	Internal flying control locks.	Removed and stowed.
88.	Pilot's seat.	Adjust for height.
89.	Rudder pedals.	Adjust for length.
90.	Flying controls.	Full and correct movement.

Cockpit checks.

N.B.—Work from left to right and then down the centre.

	ITEM	CHECK
91.	Port window and D.V. panel.	Operation.
92.	Bomb door selector lever.	Up.
93.	Air intake control.	Cold air.
94.	Fuel jettison control.	Normal.
95.	Navigation lights switch.	As required.
96.	Auto pilot main switch (Mk. 4 auto pilot only).	Off.
97.	Auto pilot control cock.	Spin.
98.	Auto pilot clutches.	In. Engage controls.
99.	Windscreen de-icing pump.	Operation.
100.	Magnetic compass.	Serviceability.
101.	External lights master switch.	As required.
102.	U/c position indicator.	Switch on. (Mks. 1, 3 and 7). Operation.
103.	Direction indicator.	Cage.
104.	Identification lights.	As required.
105.	D.R. compass.	Off.
106.	Boost control cut-out lever.	Off.
107.	No. 1 and No. 2 engine master cocks.	Off.

	ITEM	CHECK
108.	Landing lamps.	Operation. Retracted.
109.	D.R. compass repeater.	Synchronise.
110.	Auto pilot master switch (Mk. 8 auto pilot only).	Off.
111.	Ignition switches.	Off.
112.	Boost gauges.	Static readings.
113.	Flap indicator.	Switch on. Reading against position of flaps.
114.	Super-charger switch.	Low gear.
115.	No. 3 and No. 4 engine master cocks.	Off.
116.	Fuel cut-off switches (Mks. 3 and 10 air-craft).	Engine on.
117.	Pneumatic pressure gauge.	Available pressure. Pressure at each wheel brake.
118.	Radiator shutters.	Automatic.
119.	Starboard window and D.V. panel.	Operation.
120.	Throttle control friction adjuster.	Function.
121.	R.p.m. control friction adjuster.	Function.

	ITEM	ITEM
122.	Elevator trimmer.	Full and correct movement.
123.	Aileron trimmer.	Full and correct movement.
124.	Rudder trimmer.	Full and correct movement.
125.	Pilot's harness.	Adjust. Test lock.
126.	Intercom.	Adjust headset. Test with crew.
127.	Call lights.	Test with crew.
128.	Oxygen.	Delivery.
129.	Entrance ladder.	Stowed.
130.	Entrance door.	Secured.
131.	Ground/ flight switch.	Ground (if necessary).

Start and warm up the engines (see para. 38).

	ITEM	ITEM
132.	Ground/ flight switch.	Flight.
133.	Engines.	Set to 1,200 r.p.m.
134.	Radiator shutters.	Open.
135.	D.R. compass.	On and setting.
136.	Fuel flow.	Check on all tanks with booster pumps on and off.
137.	Booster pumps.	Off.
138.	Vacuum pumps.	Changeover cock. Suction gauge readings.
139.	Flaps.	Operation. Return selector to neutral.

	ITEM	CHECK
140.	Pneumatic supply.	Pressure increasing to maximum.
141.	Radio.	Test V.H.F. and other radio aids. Altimeter setting with control.
142.	Altimeter.	Set.
143.	Direction indicator.	Set with magnetic compass. Compare with D.R. compass. Uncage.
144.	Engine temperatures and pressures.	Within limits.
145.	Fuel flowmeters (if fitted).	Operation.
146.	Generators.	Check during run-up.

Run up and test the engines (see para. 39).

Pilot's checks before and during taxying.

	ITEM	CHECK
147.	All hatches.	Secure.
148.	Bomb doors.	Closed.
149.	D.R. compass.	Normal.
150.	Chocks.	Clear.
151.	Taxying.	As soon as possible test brakes. Direction indicator for accuracy. Artificial horizon for accuracy. Brake pressure. Pressure head heater on if required.

Checks before take-off.

	ITEM	CHECK
152.	Trimming tabs—	
	Elevator.	2 divs. nose heavy.
	Rudder.	Neutral.
	Aileron.	Neutral.
153.	Throttle and r.p.m. controls friction dampers.	Tighten.
154.	Superchargers.	Low gear.
155.	Air intake control.	As required.
156.	R.p.m. control levers.	Maximum r.p.m.
157.	Fuel.	Contents. Master engine. cocks on. Required tanks selected. Booster pumps on in Nos. 1 and 2 tanks. Fuel cut-off RUN (Mks. 3 and 10). Crossfeed cock off.
158.	Flaps.	20° down.
159.	Radiator shutters.	Automatic.
160.	Direction indicator.	Set to magnetic compass and uncaged.
161.	Auto pilot.	Clutches in. Control cock spin.
162.	Engines.	Clear.
163.	Harness.	Adjusted and locked.
164.	Crew.	Warn.

Checks in flight as necessary.

Checks before landing.

When entering the circuit :—

	ITEM	CHECK
165.	Crew.	Warn.
166.	Auto pilot.	Control cock spin.
167.	Superchargers.	Low gear.

ITEM	CHECK
168. Air intake control.	As required.
169. Pneumatic supply.	Pressure sufficient. Delivery to each wheel brake.
170. Fuel.	Contents. Select fullest tanks. Booster pumps. on in Nos. 1 and 2 tanks. If either No. 1 or No. 2 tanks are empty keep their booster pumps off.

Then reduce speed to 150 knots and check :—

ITEM	CHECK
171. Flaps.	20° down.
172. Undercarriage.	Down and locked (green lights on).
173. R.p.m. control levers.	Set as required. 2,850 r.p.m. on final approach.
174. Flaps.	As required on final approach.
175. Harness.	Locked.

Checks after landing.

When clear of landing area :—

ITEM	CHECK
176. Radiator shutters.	Open.
177. Flaps.	Up. Selector neutral.
178. R.p.m. control levers.	Set to max. r.p.m. position.
179. Booster pumps.	Off.
180. Pressure head heater.	Off if necessary.
181. Brake pressure.	Sufficient for taxying.

On reaching dispersal.

Idle the engines at 800-1,000 r.p.m. for a short period, test each magneto for a dead cut, then turn off the engine master cocks or operate the fuel cut-off switches or pushbuttons as applicable and when the engines have stopped :—

ITEM	CHECK
182. Ignition switches.	Off.
183. Undercarriage indicator.	Off. (Mks. 1, 3 and 7).
184. All fuel cocks.	Off.
185. Flaps.	Select down. Indicator off. (Mks. 1, 3 and 7).
186. Electrical services.	All off.
187. Direction indicator.	Caged.
188. D.R. compass.	Off.
189. Fuel cut-off switches (Mks. 3 and 10 aircraft.)	Run.
190. Chocks.	In position.
191. Brakes.	Off.
192. Flying controls.	Locked
193. Radiator shutters.	Automatic.
194. Ground/flight switch	Ground.
195. Intercom.	Off.
196. Hydraulic accumulator.	Reading 220 lb./sq. in.
197. Nitrogen main cock (if fitted).	Off.
198. Oxygen main cock.	Off.
199. Static vents.	Plugs in.
200. Pressure head.	Cover on.
201. Form 700.	Sign if necessary.
202. Authorisation book.	Sign.

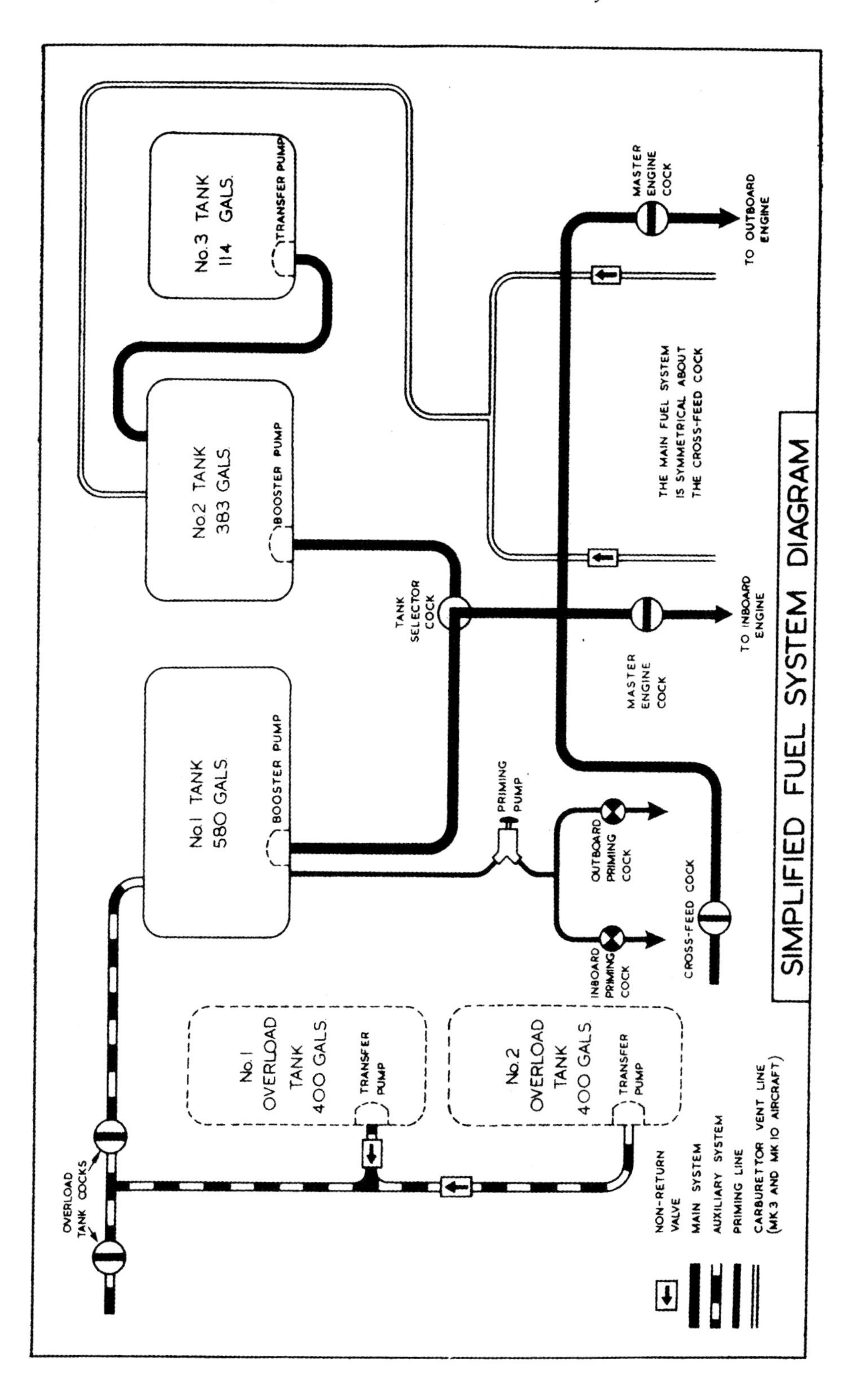
SIMPLIFIED FUEL SYSTEM DIAGRAM
No.3 TANK
114 GALS.
TRANSFER PUMP
No.2 TANK
383 GALS.
BOOSTER PUMP
No.1 TANK
580 GALS.
BOOSTER PUMP
No.1
OVERLOAD
TANK
400 GALS.
TRANSFER PUMP
No.2
OVERLOAD
TANK
400 GALS.
TRANSFER PUMP
OVERLOAD TANK COCKS
TANK SELECTOR COCK
PRIMING PUMP
OUTBOARD PRIMING COCK
INBOARD PRIMING COCK
CROSS-FEED COCK
MASTER ENGINE COCK
TO OUTBOARD ENGINE
MASTER ENGINE COCK
TO INBOARD ENGINE
THE MAIN FUEL SYSTEM IS SYMMETRICAL ABOUT THE CROSS-FEED COCK
NON-RETURN VALVE
MAIN SYSTEM
AUXILIARY SYSTEM
PRIMING LINE
CARBURETTOR VENT LINE (MK 3 AND MK 10 AIRCRAFT)

A.P. 2062 A, C, F, H, K, L & M—P.N.
Pilot's Notes

PART I
DESCRIPTIVE

NOTE.—Throughout this publication the following conventions apply :—

(a) The numbers quoted in brackets after items in the text refer to the illustrations in Part V.

(b) Words in capital letters indicate the actual markings on the control concerned.

(c) Unless otherwise stated, all speeds quoted are indicated airspeeds.

INTRODUCTION

The main differences between the Lancaster Mks. 1, 3, 7 and 10 are in the power plants.
The Lancaster 1 is fitted with Merlin 20, 22 or 24 engines which have S.U. float-type carburettors.
The Lancaster 7 is fitted with Merlin 24 engines, and an electrically-operated mid upper turret.
The Lancaster 3 and 10 are fitted with Merlin 28, 38, or 224 engines which have Bendix-Stromberg pressure-injection carburettors. Lancaster 10 aircraft are Canadian built and differ from the British-built Lancasters in some of the instruments and in the electrical system.
Hydromatic 3-bladed propellers are fitted to all marks.

FUEL AND OIL SYSTEMS

1. Fuel tanks

(i) Three self-sealing fuel tanks are fitted in each wing, numbered 1, 2 and 3 outboard from the fuselage. The capacities are :—

Port and Starboard No. 1 ...	580 gallons each
Port and Starboard No. 2 ...	383 gallons each
Port and Starboard No. 3 ...	114 gallons each
	1,077 gallons each side
	2,154 gallons in all

PART I—DESCRIPTIVE

The fuel from Nos. 1 and 2 tanks in each wing feeds through a tank selector cock directly to the engines on that side. The fuel in each No. 3 tank is transferred, when desired, into the corresponding No. 2 tank by a transfer pump. The fuel systems in each wing are independent, but are connected by a cross-feed pipe and cock.

(ii) Provision is made on some aircraft for carrying one or two 400-gallon tanks fitted in the bomb bay ; these tanks are connected so that their contents may be transferred into either or both No. 1 tanks and thence to the engines.

(iii) When the maximum bomb load is to be carried, the No. 2 tanks should be filled first, and the remainder of the fuel put in No. 1 tanks. This is on account of strength considerations of the aircraft structure.

2. **Fuel cocks**

(i) Four engine master cocks, (24) and (30), are fitted on the lower centre of the front panel, two on either side of the throttle quadrant. On Mks. 1 and 7 aircraft the engine master cocks also operate the slow-running cut-outs.

(ii) Two tank selector cocks (77) are fitted on the engineer's panel and control the fuel from Nos. 1 and 2 tanks in each wing.

(iii) A cross-feed cock, marked BALANCE COCK, is on the floor just forward of the front spar. It is reached through a hole in the spar cover and interconnects the port and starboard systems.

(iv) When the 400-gallon tanks are fitted in the bomb bay, they each have an ON-OFF cock situated behind the front spar in the centre of the fuselage.

3. **Vapour vent system** (Mks. 3 and 10 aircraft only).

A vent pipe from the two carburettors on one side of the aircraft is connected to the No. 2 tank on the same side, and allows vapour and a small quantity of fuel (approx. ½ gal. per hour, per carburettor, but some carburettors may have a second vent allowing up to 10 gallons per hour) to return to the tank. This type of carburettor is designed to work full of fuel, and it therefore requires the vent to carry away any petrol vapour and dissolved

PART I—DESCRIPTIVE

air. The vent also assists in re-establishing the flow of fuel to the carburettors when the pipe-lines and pump have been run dry due to a tank emptying.

4. **Nitrogen system**

On some aircraft provision is made to allow nitrogen to be fed into the fuel tanks, as fuel is used, to avoid the risk of explosion should they be holed in flight. The control cock is on the starboard side of the fuselage between the front and rear spars and when it is required to use the system the cock should be turned fully on before take-off, and turned off after landing. A pressure gauge is fitted above the control cock.

5. **Electric fuel booster and transfer pumps**

A booster pump is fitted in each No. 1 and No. 2 tank, and a transfer pump is fitted in each No. 3 tank which is used to replenish the corresponding No. 2 tank by switching on the No. 3 tank pump. The pumps are controlled by switches (78) mounted on the engineer's panel; the switches either have three positions, the down position being marked ON, the centre position OFF and the up position TEST, or else they have two positions, OFF and ON and a test pushbutton is mounted immediately above each switch. The TEST position of the switch (or the test pushbutton) is used in conjunction with an ammeter (68) on the engineer's panel for checking the pump operation. Similar switches control the pumps in the overload tanks (when fitted) to transfer their contents to the No. 1 tanks. The No. 3 tank switches are protected by guard covers, to prevent inadvertent operation.

6. **Fuel gauges**

(i) On Mks. 1, 3 and 7 aircraft, the switch (76) on the engineer's panel must be set ON before the fuel contents gauges (74) will indicate. On Mk. 10 aircraft there is no fuel contents gauge switch; the gauges will indicate whenever electrical power is available.

PART I—DESCRIPTIVE

(ii) Mods. 1198 and 1384 introduce "gallons gone" fuel flowmeters for each engine. Two combined indicators are mounted on the engineer's panel, and the counters can be set to zero by rotating anti-clockwise the knurled nut on the left-hand side of each indicator.

7. **Fuel pressure indicators**

Fuel pressure warning lights (79) on Mks. 1, 3 and 7 aircraft show when the fuel pressure at the carburettor falls appreciably below normal. They are switched off by the fuel contents gauges switch and this switch must, therefore, always be on in flight. On Mk. 10 aircraft, fuel pressure gauges (73) are fitted on the engineer's panel and indicate whenever electrical power is available.

8. **Priming system**

There is one cylinder priming pump in each inboard engine nacelle, drawing fuel from the No. 1 tank on that side ; each pump serves one inboard and one outboard engine. On Mks. 1, 3 and 7 aircraft this is accomplished by having two priming cocks fitted in each nacelle. On Mk. 10 aircraft the priming pump handle is turned to the left to prime the left engine, to the right to prime the right engine, and to the mid-position for off. Another cock and a short pipe may be fitted beside the priming pump and can be used to connect an outside supply of high volatility fuel for cold weather starting.

9. **Fuel jettisoning**

Originally fuel was able to be jettisoned if required from No. 1 tanks, and the control is on the floor to the left of the pilot's seat. This jettisoning system is now inoperative and the control should be kept in the normal (fully clockwise) position.

10. **Oil system**

Each engine has its own oil tank. The tanks, which are self-sealing, have a capacity of 37½ gallons of oil with 4½ gallons air space. An oil dilution system is fitted and is operated by pushbuttons (81) on the engineer's panel.

PART I—DESCRIPTIVE

MAIN SERVICES

11. Hydraulic system

(i) Each turret is operated by an individual engine-driven pump:

No. 1 engine	Tail turret
No. 3 engine	Front turret
No. 4 engine	Mid-upper turret (except on Mk. 7 aircraft where the mid-upper turret is electrically operated).

(ii) Two pumps (one each on No. 2 and No. 3 engines) operate the following services through a small accumulator:—

Air-intake shutters.
Bomb-doors.
Flaps.
Undercarriage.

(iii) A handpump for ground-test purposes is mounted in the fuselage, but owing to its small capacity, it is impracticable for use in flight.

(iv) The accumulator has an air charging valve and a pressure gauge which should read 220 lb./sq. in. when there is no hydraulic pressure in the system; misleading pressure gauge readings occur if the accumulator air pressure is incorrect. The gauge should read between 800-900 lb./sq. in. under working pressure when the cut-out operates, isolating the pumps. The accumulator then provides the initial pressure to operate the various systems. When the pressure falls, the pumps will automatically be cut in to operate the system and build up accumulator pressure again.

12. Pneumatic system

(i) A compressor on No. 3 engine charges an air bottle and operates:—

Wheel brakes.
Radiator shutters.
Supercharger rams.
Fuel cut-off rams (on Mks. 3 and 10 aircraft only).

When Lincoln type undercarriage is fitted (Mod. 1195)

PART I—DESCRIPTIVE

the air bottle is charged to 450 lb./sq. in. ; otherwise to 300 lb./sq. in.
A pressure-maintaining valve in the supply line from the air bottle only allows pressure to be supplied to the radiator shutters, superchargers and fuel cut-off rams, if the pressure in the air bottle exceeds 160 lb./sq. in. This is to ensure sufficient pressure for the brakes which operate at 125 lb./sq. in. It is necessary therefore to check on the pneumatic pressure gauge that the pressure is sufficient before high gear is engaged, or radiator override switches or fuel cut-off controls are operated.

(ii) A compressor on No. 2 engine operates the automatic pilot and the Mk. 14 bombsight. For the operation of the bombsight, the automatic pilot control cock must be set to OUT except on those aircraft in which Mod. 1161 is incorporated.

13. **Vacuum system**

A vacuum pump is fitted on No. 2 and No. 3 engines, one for operating the instruments on the instrument panel, and the other for operating the gyros of the Mk. 14 bombsight. The change-over cock (17) is on the right of the front panel beside the suction gauge (23), and in the event of failure of the vacuum pump supplying the flying instruments, the change-over cock can be used to connect the serviceable pump with the flying instruments, and cut out the bombsight. It is not possible to operate both the flying instruments and the bombsight on one vacuum pump.

14. **Electrical system**

(i) Two 1,500 watt generators are fitted on all aircraft. except Mk. 7 which have two 3,000 watt generators.
The generators are fitted one on No. 2 and one on No. 3 engine and, connected in parallel, charge the aircraft batteries (24 volt) and supply the usual lighting and other services including :—

Auto pilot.
Bomb gear and bombsight.
Camera.
Dinghy release.
D.R. compass.

PART I—DESCRIPTIVE

Electro-pneumatic rams for :—
Radiator shutters.
Supercharger gear change.
Air-intake emergency heat control.
Fuel cut-offs (on Mk. 3 and Mk. 10 aircraft).
Engine starters and booster coils.
Fire-extinguishers.
Flaps position indicator.
Fuel booster and transfer pumps.
Fuel contents and flowmeter gauges.
Mid-upper turret (on Mk. 7 aircraft).
Oil dilution.
Propeller feathering.
Pressure-head heater.
Radio and radar.
Undercarriage position indicator.

An alternator may be fitted to Nos. 1 and 4 engines, to supply the radar equipment.

(ii) A ground/flight switch on the starboard side of the fuselage immediately aft of the front spar is used to isolate the aircraft batteries when the aircraft is parked or when using a ground starter battery.

(iii) The electrical control panel is fitted on the starboard side of the fuselage forward of the front spar, and mounted on the panel are a voltmeter, two generator switches, two ammeters and two earth warning lights. On Mk. 10 aircraft overload switches are fitted on the heavier circuits and flick off if the current becomes too great. If a switch flicks off it should be re-set only after the overloaded circuit has been allowed to cool for 30 seconds. If the switch flicks off again it indicates that the circuit is defective.

AIRCRAFT CONTROLS

15. **Flying controls**

The flying controls are conventional, and each rudder pedal is adjustable by holding aside the spring-loaded latch on each inside pedal arm and raising and moving the foot-rest over the ratchet mechanism.

PART I—DESCRIPTIVE

16. **Flying controls locking gear**

The controls locking gear is stowed on the starboard side of the fuselage between the main spars, and consists of :—

(a) a strut to be fastened to the top of the pilot's seat and to a bracket on the control column.

(b) a strut to be inserted at one end into the cockpit left-hand rail, and fitted by two screwed hooks to the handwheel to prevent it rotating.

(c) a T-tube with a transverse member, to be inserted into the hollow footrest of each rudder pedal, and the other end attached to the bracket on the control column.

17. **Trimming tabs**

The elevator (62), rudder (63) and aileron (61) tab controls, on the right of the pilot's seat, all operate in the natural sense and each has an adjacent indicator showing the setting of the tab.

18. **Wheel brakes**

The brakes are applied by operating the lever (56) on the control column, and differential action is obtained by a relay valve connected to the rudder bar. The brakes may be locked for parking by compressing the lever fully and engaging the locking catch which is beside the lever. The available pneumatic pressure, and the pressure at each brake is shown on a triple pressure gauge (20) mounted on the right of the front panel.

19. **Undercarriage control**

The undercarriage lever (64) is locked in the DOWN position by a safety bolt (65) which has to be held aside in order to raise the lever. The bolt engages automatically when the lever is set to DOWN. The undercarriage may be lowered in an emergency by compressed air (see para. 61). There is no automatic lock other than the safety bolt, to prevent the undercarriage being raised by mistake when the aircraft is on the ground.

PART I—DESCRIPTIVE

20. Undercarriage indicator

On Mks. 1, 3 and 7 aircraft, the indicator (39) shows as follows :—

Undercarriage locked down	...	Two green lights
,, unlocked ...	...	Two red lights
,, locked up ...	...	No lights

The indicator switch (4) is interlocked so that it must be on when the No. 1 and No. 2 engine ignition switches are on. An auxiliary set of green lights can be brought into operation by pressing the central knob if failure of the main set is suspected. The red lamps are duplicated so that failure of one lamp does not affect the indication of undercarriage unlocked. The lights can be dimmed by turning the central knob.

On Mk. 10 aircraft a pictorial type of indicator is fitted. When the indicator switch is on and electrical power is available, the pictorial indicator shows the position of the undercarriage wheels and wing flaps. The disappearance of small red flags shows when the wheels are locked up or down.

21. Undercarriage warning horn

The horn sounds if either inboard throttle is closed when the undercarriage is not locked down. The outboard throttles do not operate the horn. A testing pushbutton and lamp are behind the pilot's seat on the cockpit port rail.

22. Flaps control

If the flaps have been selected partly down, and it is desired to lower them fully, it may be found that they will not lower further. This is due to the pressure in the accumulator having fallen below the pressure required to operate the flaps against the external air pressure but not sufficiently to cause the hydraulic pumps to cut in. To overcome this, move the selector (60) to UP, and then immediately put it fully DOWN ; it may be necessary to repeat this process more than once. This causes the hydraulic pumps to cut in. After the flaps have been selected fully down for landing, the selector should be left DOWN until landing is complete, to avoid any possibility of the flaps creeping up.

PART I—DESCRIPTIVE

On Mks. 1, 3 and 7 aircraft, the position indicator (26) is switched on by a separate switch (27).
In an emergency the flaps may be lowered by compressed air after lowering the undercarriage (see para. 62).

23. **Bomb doors**

The control (43) has two positions only. The bomb release system is rendered operative soon after the doors begin to open and before they are fully open. The position of the doors must therefore be checked visually before releasing bombs. If the bomb doors open only part way and then stop, it is probably due to icing around the hinges and joints, which raises the hydraulic pressure sufficiently to bring the cut-out into operation, stopping any further movement of the doors. If the bomb doors selector is moved to SHUT and then immediately to OPEN, the doors will usually open further ; it may be necessary to repeat this several times to get the doors fully open. As strenuous pumping for 15 minutes is required to open the doors with engines stopped, they should be opened before stopping engines if the aircraft is to be bombed up before the next flight, or if necessary for servicing purposes.

24. **Automatic pilot**

On early aircraft the Mk. 4, and on later aircraft the Mk. 8 automatic pilot, is fitted, driven by a compressor in No. 2 engine.

ENGINE CONTROLS

25. **Throttle controls**

(i) *Merlin* 20, 22, 28 *and* 38 *engines.* Climbing boost, +9 lb./sq. in., is obtained with the throttle levers (28) at the gate. On Merlin 20 installations, going through the gate gives +12 lb./sq. in. at ground level only ; on Merlin 22, 28 and 38 engines it gives +14 lb./sq. in. When the boost control cut-out (32) is pulled, + 14 lb./sq. in. is obtainable in low gear, and +16 lb./sq. in. in high gear on all the above Merlins.

(ii) *Merlin* 24 *and* 224 *engines.* The throttle quadrant is fitted with a gate at +9 lb./sq. in boost ; the fully-forward

PART I—DESCRIPTIVE

position gives +14 lb./sq. in. at ground level only. When the boost control cut-out is pulled, +18 lb./sq. in. is obtainable in either gear.

26. **Mixture control**

(i) *Merlin* 20, 22 *and* 24 *engines.* (Mks. 1 and 7 aircraft.) S.U. float-type carburettors are fitted. The mixture strength is automatically controlled by boost pressure, and the pilot has no separate mixture control. A weak mixture is obtained below + 7 lb. /sq. in. (+4 lb./sq. in. on Merlin 20). The carburettor slow-running cut-outs are operated by closing the engine master cocks.

(ii) *Merlin* 28, 38 *and* 224 *engines.* (Mks. 3 and 10 aircraft.) Bendix-Stromberg pressure injection carburettors are fitted. There is no pilot's mixture control, the mixture strength being regulated by the boost so that an economical mixture is obtained below +7 lb./sq. in. The fuel cut-offs, which are used in starting and for stopping the engines, are operated by electro-pneumatic rams controlled by four two-position switches (11) or four push-buttons (if Mod. 1753 is fitted) mounted on the pilot's panel above the engine starter buttons. When two-position switches are fitted the top position is the ENGINE RUN position, and the bottom position is the IDLE-CUT-OFF position ; when pushbuttons are fitted, they have to be held in to keep them in " cut-off."

NOTE.—(a) If the pneumatic supply pressure is less than 160 lb./sq. in., it is possible to start the engines with the fuel cut-off switches in the IDLE-CUT-OFF position ; then, when the supply pressure builds up the cut-off rams will operate and all four engines will stop.

(b) When the aircraft is parked the fuel-cut-offs should be left in the ENGINE RUN position. If left in the IDLE-CUT-OFF position the rams will return to the " engine run" position when the electrical current is switched off, and back to " cut-off " when it is switched on again, with a consequent waste of pneumatic pressure.

PART I—DESCRIPTIVE

27. Propeller controls

(i) The r.p.m. control levers (29) are mounted in a quadrant on the engine controls pedestal and vary the governed r.p.m. from 3,000 down to 1,800.

(ii) The feathering pushbuttons (19) are mounted on the right of the front panel and when Mods. 1067 and 1314 are fitted, engine fire warning lights are mounted in the respective feathering pushbuttons (see paras. 54 and 69).

28. Superchargers control

The superchargers gear-change is electro-pneumatically operated by a switch on the front panel immediately below the engine speed indicators. The switch, which has two positions, MS and FS, controls all four engines simultaneously and a red warning light (25) beside it indicates if FS gear is engaged when the main wheels are down. In the event of electrical or pneumatic failure, the rams will stay at, or return to, the MS gear position.

29. Carburettor air-intake heat control

A single lever for the hydraulic operation of all four carburettor's warm air intakes is on the left of the pilot's seat.

When Mod. 1198 is fitted the control has 3 positions, COLD, WARM and HOT (emergency). If the control is at WARM, air is then drawn into each engine through the warm air-intake inside the bottom engine cowling. When moved to HOT, a flap in the bottom engine cowling is opened by an electro-pneumatic ram and allows hot air to be drawn from the radiator to the engine. The WARM position may be used to prevent the formation of ice when flying in icing conditions, but this will reduce the range (see para. 52 (iii)). If, instead, the flight is continued in COLD until carburettor icing becomes evident, the HOT position should then be used, but only until the ice has cleared.

If Mod. 1198 is not fitted, there are only two positions COLD and WARM, but the latter position should be regarded as having a similar function to the HOT position when the 3-position control is fitted, as air is drawn from the radiator to the engine through a hole in the engine cowling when the control is at WARM.

PART I—DESCRIPTIVE

30. Radiator shutters

The radiator shutters are controlled by two-position switches marked AUTOMATIC and OPEN, and are mounted on the cockpit right-hand wall. The shutters are automatically operated by a thermostat when they are set to the AUTOMATIC position ; when the switches are in the OPEN position the thermostatic control is overriden, and this position should be used for all ground running and taxying.
When the aircraft is parked, the shutters should be left in the AUTOMATIC position. If left in the OPEN position, they will close when the electrical current is switched off, and open when it is switched on again, with a consequent waste of pneumatic pressure.

31. Ignition and starting controls

The ignition switches (7), booster-coil master switch (10) and the starter pushbuttons (14) are all at the top centre of the front panel. The starter pushbuttons operate the booster coils when the booster-coil switch is ON.

OTHER EQUIPMENT

32. Cockpit lighting

(i) Lighting is provided by two floodlights, which are controlled by adjacent dimmer switches in the cockpit roof. The engineer's panel is provided with a separate light and dimmer switch.

(ii) On some aircraft a single emergency light is fitted for use only in the event of complete electrical failure. It is powered by a separate small battery and is controlled by a switch on the left-hand side of the front panel.

33. Cockpit heating and ventilation

The cockpit is heated by warm air from the two radiators mounted in the mainplane leading edge and connected to the inboard engine cooling systems. On each side of the fuselage, just forward of the front spar, is a control knob which operates a shutter in the air duct. When turned anti-clockwise it opens the inlet to the cabin and closes the by-pass to the outer air. To control the escape

LANCASTER

FINAL CHECKS FOR TAKE-OFF

TRIM	... ELEVATORS : 2 DIVS. NOSE HEAVY
	RUDDER : NEUTRAL
	AILERON : NEUTRAL
PROPELLERS	... MAX. R.P.M.
FUEL	... MASTER COCKS ON
	CORRECT TANKS SELECTED
	NOS. 1 AND 2 TANKS BOOSTER PUMPS ON
	CROSSFEED COCK OFF
	FUEL CUT-OFFS "RUN"
FLAPS	... 20° DOWN
SUPER-CHARGERS	... LOW GEAR
AUTO-PILOT	... SPIN

FINAL CHECKS FOR LANDING

FUEL ...	CHECK CONTENT[illegible]
	CORRECT TAN[illegible] S[illegible]CTED
	NOS. I AND [illegible]ANKS BOOSTE[illegible]MPS ON
SUPER-CHARGERS	LOW GEAR
BRAKES ...	OFF CHECK PRESSURES
UNDER-CARRIAGE ...	DOWN AND LOCKED
PROPELLERS ...	2,850 R.P.M. ON FINAL
FLAPS ...	AS REQUIRED

PART I—DESCRIPTIVE

of the air from the cabin an extractor louvre is provided on each side of the nose of the fuselage.

34. **Pilots' seats**

The left-hand pilot's seat is provided with hinged arm-rests and is adjustable for height by a lever on the left-hand side. The right-hand pilot's seat is collapsible and folds up against the side of the fuselage, being secured by a strap to the cockpit rail when not in use. A back-rest is positioned above the seat and can be folded up against the fuselage side when not required. A bar is provided as a foot-rest and, when not in use, slides under the left seat platform.

35. **Windscreen de-icing**

Two de-icing sprays for the windscreen are operated by a handpump (59) on the floor, forward and to the left of the pilot's seat. The pump is operated by pressing down the handle which is spring-loaded to return at a pre-set speed according to the setting of the needle valve of the pump. A setting of $1\frac{2}{3}$ turns on the needle valve is recommended when the pump, operated approximately once a minute, will give a delivery at the rate of two pints per hour. The tank, which also supplies the bomb-aimers' panel, is of approximately 4 gallons capacity and is fitted on the starboard side between the nose and the front centre section of the aircraft.

36. **Oxygen system**

The pilot's flexible oxygen pipe (51) is secured by spring clips to the cockpit left-hand rail, and the economiser is below the rear end of the pilot's floor. A Mk. 8 series regulator (18) which controls the supply throughout the aircraft is fitted on the right of the front panel.

PART II

HANDLING

37. Management of the fuel system

(i) *Use of tanks.* Structural considerations render it advisable that fuel should be kept outboard as much as possible.

(a) Start and warm up on No. 1 tanks. Run up and take-off on No. 2 tanks, and continue to fly on these tanks for the first hour of flight. This allows space for carburettor venting, where applicable (see para. 3).

(b) After the first hour of flight No. 1 tanks should be selected. When nearly empty, transfer the fuel from the fuselage overload tanks, if in use, by turning on both long-range fuel cocks (behind the front spar) and switching on the overload tank pump switches. The fuel contents gauges switch must also be on. Transfer of fuel from the long-range tanks takes approximately one hour. Turn off each long-range tank cock and the appropriate pump switch when the tank is empty.

(c) Continue to run on No. 1 tanks until empty ; then re-select No. 2 tanks and use until approximately 200 gallons remain in each. Then transfer the contents of No. 3 tanks by switching on the transfer pumps.
Switch off the pumps when No. 3 tanks are empty.

(ii) *Use of the booster pumps*
The main use of the booster pumps in No. 1 and No. 2 tanks is to maintain fuel pressure at altitudes of approximately 17,000 ft. and over in temperate climates, but they are also used for raising the fuel pressure before starting and to assist in re-starting an engine during flight. If one engine fails and the booster pump is not ON, air may be drawn back into the main fuel system before the engine master cock of the failed engine can be closed, thus causing the failure of the other engine on the same

side. At take-off, therefore, the pumps in Nos. 1 and 2 tanks must be switched on ; this is also a precaution against fuel failure during take-off as an immediate supply is available by changing over the tank selector cock. The pump in each tank in use should also be switched on at any time when a drop in fuel pressure is indicated or when it is necessary to run all engines from one tank by opening the cross-feed cock.

(iii) *Testing the booster and transfer pumps*

Before starting the engines, each booster and transfer pump should be tested by the ammeter fitted on the engineer's panel ; to do this the switch for each pump (on Mks. 1, 3 and 10 aircraft) should in turn be set to the up (TEST) position, after ensuring that the engine master cocks are OFF. On Mk. 7 aircraft the pumps are tested by pressing the test pushbuttons above the switches. The ammeter reading should in all cases be perfectly steady and should be between 4 and 7 amps. on Mks. 1, 3 and 7 aircraft, and between 3 and 5 amps. on Mk. 10 aircraft.

(iv) *Use of cross-feed cock*

The cock should be closed at all times, unless it is necessary in an emergency to feed fuel from the tanks in one wing to the engine(s) in the other wing. If the cross-feed cock is open for this purpose, select the tank from which it is desired to cross-feed fuel, switch on the pump in this tank, and turn off the selector cock of the opposite wing tanks.

38. **Starting the engines and warming up**

NOTE.—On Mks. 3 and 10 aircraft the fuel booster pumps must never be switched on with the engine master cock open and the engine stationary, unless the fuel cut-off switch, or pushbutton, is in the IDLE-CUT-OFF position, and the pneumatic supply pressure not less than 160 lb./sq. in.

(i) *Preliminaries.* After carrying out the external, internal and cockpit checks, laid down in the Pilot's Check List, ensure :—

PART II—HANDLING

Engine master cocks ...	OFF
Fuel cut-off controls ...	ENGINE RUN
Throttles	½ inch open
R.p.m. controls	Control levers max. r.p.m.
Superchargers control ...	MS gear (warning light out)
Air intake heat control ...	COLD
Radiator shutters... ...	Override switches at AUTOMATIC
Booster pumps	OFF
Booster-coil master switch	ON

(ii) Turn on the engine master cock of the engine to be started, and if the aircraft is to be started from external batteries, set the ground/flight switch to GROUND.

(iii) Prime the carburettor of the engine to be started by putting the fuel cut-off to IDLE-CUT-OFF (on Mks. 3 and 10 aircraft) and switching the booster pump in the No. 1 tank on for a period of 10 seconds. Switch off the booster pump and then return the fuel cut-off to the ENGINE RUN position.

(iv) High volatility fuel should be used if an outside priming connection is fitted, for priming at air temperatures below freezing. The ground crew should work the priming pump until the fuel reaches the priming nozzles ; this may be judged by an increase in resistance.

(v) Switch on the ignition and press the starter button. Turning periods must not exceed 20 seconds with a 30 second wait between each. The ground crew should work the priming pump as firmly as possible while the engine is being turned.

(vi) It will probably be necessary to continue priming after the engine has fired, and until it picks up on the carburettor. When the engine is running smoothly, proceed to prime the carburettors and start the other engines in turn.

(vii) When all the engines are running satisfactorily, switch off the booster-coil master switch. The ground crew should screw down the priming pumps and turn off the priming cocks (if fitted).

(viii) Ensure that the ground/flight switch is turned to FLIGHT and have the external battery removed if used.

(ix) Open up each engine slowly to 1,200 r.p.m. and warm up at this speed.

PART II—HANDLING

(x) While warming up carry out the checks detailed in the Pilot's Check List, items 132 to 146.

39. Exercising and Testing

(i) After warming up until the oil temperature is +15°C. and the coolant temperature is +40°C., switch the radiator shutters over-ride switches to OPEN, turn the tank selector cocks to No. 2 tanks, and test each magneto as a precautionary check before increasing power further. Then for each engine in turn :—

(a) Open up to the static boost reading (0 lb./sq. in. under "standard atmosphere" conditions) and exercise and check the operation of the constant speed propeller by moving the lever over its full range at least twice. With the lever fully forward check that the r.p.m. are within 50 of those normally obtained.

(b) At the same boost check the operation of the supercharger. R.p.m. should fall, boost rise and the supercharger warning light should come on when high gear is engaged. Change back to low gear and ensure that the original conditions are restored.

(c) At the same boost check with the engineer that the generators on Nos. 2 and 3 engines are charging.

(d) At the same boost test each magneto in turn. If the single ignition drop exceeds 150 r.p.m., but there is no undue vibration, the ignition should be checked at higher power, see below ; if there is marked vibration the engine should be shut down and the cause investigated.

NOTE.—The following full power checks should be carried out after repair, inspection other than daily, when the single ignition drop at the static boost reading exceeds 150 r.p.m. or at the discretion of the pilot. Except in these circumstances if the checks above are satisfactory no useful purpose will be served by a full power check.

(e) Open the throttle fully and check take-off boost and r.p.m. This check should be as brief as possible.

(f) Throttle back until the r.p.m. fall just below the take-off figure and test each magneto in turn. If the single ignition drop exceeds 150 r.p.m. the aircraft should not be flown.

(g) After completing the checks, either at the static boost reading, or at full power, steadily move the throttle to the fully-closed position, and check the minimum idling r.p.m. then open up to between 1,000 and 1,200 r.p.m.

(ii) Before and during taxying carry out the checks in the Pilot's Check List, items 147 to 151.

40. **Taking off**

(i) After carrying out the checks, items 152 to 164 laid down in the Pilot's Check List, clear the engines by opening up to the static boost reading if the run-up has not been carried out immediately prior to taxying on to the runway.

(ii) Align the aircraft carefully on the runway, making certain that the tailwheel is straight.

(iii) Release the brakes and open the throttles slowly to the take-off position.

(iv) Keep straight by coarse use of the rudder and by differential throttle opening.

(v) As speed is gained, ease the control column forward to raise the tail. Do not attempt to raise the tail by exerting a heavy push force on the control column during the early stages of the take-off run.

(vi) At 65,000 lb. ease the aircraft off the ground at 90 knots and at 72,000 lb. at 105 knots.

(vii) When comfortably airborne, brake the wheels and retract the undercarriage.

(viii) With flaps 20° down, safety speed at 65,000 lb. is 145 knots when using +18 lb./sq. in. boost and 3,000 r.p.m.; at 72,000 lb. it is 150 knots. In view of these high speeds power should, where practicable, be reduced early after take-off.

(ix) At a safe height raise the flaps in stages. Then return the selector to neutral.

(x) The booster pumps in Nos. 1 and 2 tanks may be switched off after the initial climb, but if a warning light comes on (or on Mk. 10 aircraft the fuel pressure gauge shows less than 10 lb./sq. in.), switch on No. 2 pumps immediately.

PART II—HANDLING

41. Climbing

The speed for maximum rate of climb is 140 knots, but to improve control when climbing at full load speeds may be increased to 155 knots.

42. General flying

(i) *Stability.*—At normal loadings and speeds stability is satisfactory. At loads above 67,000 lb. there is a tendency for the aircraft to wallow. It is not advisable to attempt to correct this as use of the controls may aggravate it.

(ii) *Controls.*—The elevators are relatively light and effective, but tend to become heavy in turns. The ailerons are light and effective but become heavy at speeds over 225 knots, and also at heavy loads. The rudders also become heavy at high speeds.

(iii) *Changes of trim :—*

Undercarriage UP	Slightly nose up
Undercarriage DOWN ...	Slightly nose down
Flaps up to 20° from fully DOWN	Slightly nose down
Flaps up from 20°	Strongly nose down
Flaps down to 20°	Strongly nose up
Flaps fully DOWN from 20°	Slightly nose up
Bomb doors open	Slightly nose up

(iv) *Flying at reduced airspeeds.*—Flaps may be lowered about 20°, r.p.m. set to 2,650, and the speed reduced to about 115 knots.

43. Maximum performance

(i) *Climbing :—*
140 knots to 12,000 ft.
135 knots from 12,000 to 18,000 ft.
130 knots from 18,000 to 22,000 ft.
125 knots above 22,000 ft.
Change to high gear when boost has fallen to +6 lb./sq. in.

(ii) *Operational necessity*
Use high gear if the boost obtainable in low gear is more than 3 lb./sq. in. (4 lb./sq. in. with Merlin 24 or 224) below maximum boost.

PART II—HANDLING

44. Maximum range

(i) Climbing.—140 knots at +7 lb./sq. in. boost with Merlin 22, 24, 28, 38 or 224 (+4 lb./sq. in with Merlin 20) and 2,650 r.p.m. Change to high gear when maximum boost obtainable in low gear has fallen by 3 lb./sq. in.

(ii) Cruising (including descent) :—

(a) Fly in low gear at maximum obtainable boost not exceeding +4 lb./sq. in. with Merlin 20, +7 lb./sq. in. with Merlin 22, 24, 28, 38 or 224 obtaining the recommended airspeed by reducing r.p.m., which may be as low as 1,800 if this will give the recommended speed. Higher speeds than those recommended may be used if obtainable in low gear at the lowest possible r.p.m.

(b) The recommended speeds are :—

Fully loaded	
Up to 15,000 ft.	150 knots
At 20,000 ft. in high gear	140 knots
Lightly loaded	140 knots

(c) Engage high gear when the recommended speed cannot be maintained at 2,500 r.p.m. in low gear.

(iii) The use of warm air intakes will reduce air nautical miles per gallon considerably. On this installation there is no need to use warm air unless intake icing is indicated by a drop in boost.

45. Fuel capacity and consumptions

(i) Capacity :—

Two No. 1 tanks	1,160 gallons
Two No. 2 tanks	766 gallons
Two No. 3 tanks	228 gallons
Normal total ...	2,154 gallons
Two 400 gallon overload tanks ...	800 gallons
Overload total ...	2,954 gallons

PART II—HANDLING

(ii) Weak mixture consumptions, Merlin 20, 22 or 24 :—
The following figures are the approximate total gallons per hour and apply in low gear between 8,000 and 17,000 feet, and in high gear between 14,000 and 25,000 feet.

Boost lb./sq. in	R.p.m.		
	2,650	2,300	2,000
+7*	260*	225*	212*
+4	228	204	188
+2	212	188	172
0	192	172	150
-2	172	156	140
-4	152	136	124

* These figures do not apply to Merlin 20.

(iii) Weak mixture consumptions Merlin 28, 38 and 224 engines :—
The following figures are the approximate total gallons per hour for the aircraft and apply in low gear at 5,000 ft. and high gear at 15,000 ft. One gallon per hour should be added for every 1,000 ft. above these heights.

Boost lb./sq. in.	R.p.m.				
	2,650	2,400	2,200	2,000	1,800
+7	240	235	217	200	—
+4	216	204	196	180	—
+2	196	184	176	164	—
0	172	164	156	144	128
−2	148	140	128	124	112
−4	124	120	108	104	96

(iv) Rich mixture consumptions, Merlin 20, 22, 24.

Boost lb./sq. in.	R.p.m.	Total gallons per hour
+14	3,000	500
+12	3,000	460
+9	2,850	380
+7*	2,650*	320*

* Merlin 20 only.

PART II—HANDLING

(v) Rich mixture consumptions, Merlin 28, 38 and 224 :—

Boost lb./sq. in.	R.p.m.	Total gallons per hour
+9	2,850	420

46. Position error corrections

(i) The position error correction is −1 knot at all speeds from 120 knots upward.

(ii) When an H_2S blister is fitted, the position error corrections are as follows :—

From	110	130	160	190	knots
To	130	160	190	250	knots
Subtract	1	2	3	4	knots

(iii) On Lancaster ASR Mk. 3 the corrections are as follows :

(a) When the lifeboat is being carried, the correction is +3 knots at speeds below and +2 knots at speeds above 140 knots.

(b) When the lifeboat is not being carried the correction varies from +1 knot at 95 knots to −1 knot at 215 knots.

47. Stalling

(i) Warning of the stall is given by slight tail buffeting, which generally commences some 5 knots before the stall itself.

At the stall the nose drops gently. Recovery is straightforward and easy.

(ii) The approximate stalling speeds are :—

	At 60,000 lb.	At 65,000 lb.
Power off		
Undercarriage and flaps up	95	105
Undercarriage and flaps down	80	85
Power on (approach conditions, −2 lb./sq. in boost, 2,850 r.p.m. set)		
Undercarriage and flaps down	70–75	—

(iii) *High speed stall.*—Adequate warning of the approach of a stall in a turn is given by strong rudder and elevator

PART II—HANDLING

buffeting. At the stall the inner wing and nose drop gently together. Recovery is immediate on pushing the control column forward.

48. **Diving**

There is a strong nose-down change of trim as speed is gained in the dive. On aircraft in which Mod. 1101 or 1131 is incorporated it is possible to recover from dives to the limiting speed without the assistance of the elevator trimming tab, even if the aircraft has been trimmed into the dive. If the elevator trimming tab is used it should be applied with care, since it is powerful and sensitive. On unmodified aircraft the elevator trimming tab should never be used to assist entry, but should be used to reduce the very heavy pull force otherwise necessary for recovery.

49. **Approach and landing**

After carrying out the checks, items 165 to 175, laid down in the Pilot's Check List, the turn into wind should be made at about 115 knots, airspeed being reduced progressively, so that the airfield boundary is crossed at the following speeds :—

	At light load 45,000 lb.	At max. landing weight 60,000 lb.
Engine assisted Flaps down	95	105

50. **Mislanding and going round again**

(i) At maximum landing weight the aircraft will climb away satisfactory with the undercarriage and flaps down at maximum climbing power.

(ii) After increasing power, maintain an airspeed of 125 knots, select undercarriage up, and while the undercarriage is rising, select the flaps up to 30° and thereafter in stages, retrimming as necessary.

51. **After landing**

(i) Carry out the checks in the Pilot's Check List, items 176 to 181.

PART II—HANDLING

(ii) Before stopping the engines, open the bomb doors if required for bombing up or servicing purposes.

(iii) *Stopping engines*

(a) *Merlin* 20, 22 *and* 24
With the engines running at 800 r.p.m. turn OFF the engine master cocks and switch OFF the ignition after the engines have stopped.

(b) *Merlin* 28, 38 *and* 224
Check that the pneumatic pressure gauge reads at least 160 lb./sq. in. If not, open up No. 3 engine to increase the pneumatic pressure, and then, with the engine running at about 800 r.p.m., move the fuel cut-off to the " cut-off " position.
Do not stop the engines by turning off the engine master cocks, as this will empty the carburettors of fuel and entail trouble in any subsequent starting attempt. When all the engines have stopped, switch off the ignition and turn off the engine master cocks.

(iv) Carry out the checks in the Pilot's Check List, items 182 to 202.

(v) *Oil dilution*

The recommended dilution period for this aircraft is :—
Air temperatures above - 10°C. one minute
" " below - 10°C. two minutes

A.P. 2062 A, C, F, H, K, L & M—P.N.
Pilot's Notes

PART III

LIMITATIONS

52. **Engine data**

(i) *Merlin* 20

Engine limitations with 100/130 grade fuel :—

	Gear	R.p.m.	Boost lb./sq. in.	Temp. °C. Coolant	Oil
MAX. TAKE-OFF 5 MINS. LIMIT	Low	3,000	+12*		
MAX. CLIMBING 1 HOUR LIMIT	Low / High	2,850	+9	125	90
MAX. RICH CONTINUOUS	Low / High	2,650	+7	105	90
MAX. WEAK CONTINUOUS	Low / High	2,650	+4	105	90
OPERATIONAL NECESSITY 5 MINS. LIMIT	Low	3,000	+14	135	105
	High	3,000	+16	135	105

* +14 lb./sq. in. is obtained by operating the boost control cut-out.

OIL PRESSURE :
MINIMUM IN FLIGHT ... 30 lb./sq. in.

MINIMUM TEMPS. FOR TAKE-OFF :
OIL ... ... +15°C.
COOLANT ... ... +40°C.

(ii) *Merlin* 22, 28 *or* 38

Engine limitations with 100/130 grade fuel :—

	Gear	R.p.m.	Boost lb./sq. in.	Temp. °C. Coolant	Oil
MAX. TAKE-OFF 5 MINS. LIMIT	Low	3,000	+14		
MAX. CLIMBING 1 HOUR LIMIT	Low / High	2,850	+9	125	90
MAX. CONTINUOUS	Low / High	2,650	+7	105	90
OPERATIONAL NECESSITY 5 MINS. LIMIT	Low	3,000	+14	135	105
	High	3,000	+16	135	105

PART III—LIMITATIONS

OIL PRESSURE :

MINIMUM IN FLIGHT ... 30 lb./sq. in.

MINIMUM TEMPS. FOR TAKE-OFF :

OIL ... ... +15°C.

COOLANT ... ... +40°C.

(iii) *Merlin* 24 *or* 224

Engine limitations with 100/130 grade fuel :—

	Gear	R.p.m.	Boost lb./sq. in.	Temp. °C. Coolant	Oil
MAX. TAKE-OFF 5 MINS. LIMIT	Low	3,000	+18*		
MAX. CLIMBING 1 HOUR LIMIT	Low / High	2,850	+9	125	90
MAX. CONTINUOUS	Low / High	2,650	+7	105	90
OPERATIONAL NECESSITY 5 MINS. LIMIT	Low / High	3,000	+18*	135	105

* +18 lb./sq. in. boost must not be used below 2,850 r.p.m.

OIL PRESSURE :

MINIMUM IN FLIGHT ... 30 lb./sq. in.

MINIMUM TEMPS. FOR TAKE-OFF :

OIL ... ... +15°C.

COOLANT ... ... +40°C.

53. **Flying limitations**

(i) The aircraft is designed for manœuvres appropriate to a heavy bomber and care must be taken to avoid imposing excessive loads with the elevators in recovery from dives and in turns at high speed. Violent use of the rudder at high speeds should be avoided.

(ii) Maximum speeds in knots :—

Diving 315
Bomb doors open 315
Undercarriage down 175
Flaps down 175

PART III LIMITATIONS

(iii) Maximum weights :—

Take-off, straight flying and gentle manœuvres	63,000 lb.
,, ,, ,, ,, ,, ,,	†64,000 lb.
,, ,, ,, ,, ,, ,,	‡65,000 lb.
,, ,, ,, ,, ,, ,,	*72,000 lb.
Landing and all forms of flying	55,000 lb.
Landing	*60,000 lb.

†This weight is permitted for Lancaster ASR. Mk. 3.

‡This weight is permitted provided the following mods. are incorporated : Mod. 503 or 518, Mod. 588 or 598, Mod. 811 or SI/RDA. 600 and Mod. 1004.

*These weights are permitted if Merlin 24 or 224 power plants are fitted, paddle-bladed propellers are fitted. Lincoln type undercarriage and tyres (Mod. 1195) are fitted and special adjustments are made to the tyre and oleo leg pressures. A careful check on aircraft structure must be kept and runways only must be used.

PART IV
EMERGENCIES

54. Feathering

(i) Close the throttle.
(ii) Press the feathering pushbutton and hold it in only long enough to ensure that it stays in by itself. Then release it and check that it springs out when feathering is complete. If it does not do so, it must be pulled out by hand.
(iii) Switch off the ignition when the propeller has stopped (or nearly stopped) rotating, and turn off the engine master cock at once.
(iv) If the engine has been feathered because of fire, operate the engine fire-extinguisher as soon as the propeller has stopped turning (see also para. 69).
(v) Engine auxiliaries which will be affected by feathering :—
No. 1 engine.—Alternator for radar, rear turret hydraulic pump.
No. 2 engine.—Generator, main services hydraulic pump, compressor for automatic pilot and computor unit of Mk. 14 bombsight, No. 1 vacuum pump.
No. 3 engine.—Generator, main services hydraulic pump, front turret hydraulic pump, compressor for pneumatic system, No 2 vacuum pump.
No. 4 engine.—Alternator for radar, mid-upper turret hydraulic pump (when fitted).

55. Unfeathering

(i) Switch on the ignition, set the throttle as for starting and the r.p.m. control lever to the minimum r.p.m. position.
(ii) Check that the fuel booster pump of the tank in use is OFF, then press the feathering pushbutton. As the engine starts to turn, set the engine master cock to ON. Continue to hold the button in until r.p.m. reach 800-1,000 Check that the pushbutton springs out when released ; if it does not do so it must be pulled out by hand.
(iii) If the propeller does not return to normal constant-speed operation it must be refeathered and then unfeathered again, releasing the button at slightly higher r.p.m.

PART IV—EMERGENCIES

56. **Engine failure during take-off**

(i) With flaps 20° down safety speed at 65,000 lb. is 145 knots, using + 18 lb./sq. in. boost and 3,000 r.p.m.

(ii) If engine failure occurs below critical speed it will always be necessary to throttle back the opposite outer engine at least partially to maintain control.

(iii) With the propeller of the failed engine feathered and rudder trim applied to relieve footload it should be possible for the aircraft to climb away at 125 knots at weights up to 65,000 lb.

(iv) As soon as the undercarriage is up, raise the flaps in stages, retrimming as necessary.

57. **Handling on three engines**

The aircraft will maintain height at loads up to 65,000 lb. on any three engines at 10,000 feet and can be trimmed to fly without footload. Maintain at least 130 knots.

58. **Landing on three engines**

Lowering of flaps to 20° and of undercarriage may be carried out as normally on the circuit but further lowering of the flaps should be left until the final straight approach.

59. **Going round again on three engines**

The decision to go round again should be made before full flap is lowered. With the flaps lowered 20° and the undercarriage down, power should be increased to +9 lb./sq. in. boost, 2,850 r.p.m. The aircraft can be controlled comfortably at 130 knots. Select undercarriage up and while it is rising, select flaps up in stages, retrimming as necessary.

60. **Flying on asymmetric power on two engines**

(i) *In flight.*—It should be possible to maintain height below 10,000 feet at 125 knots on any two engines after release of bombs and with half fuel used, but with two engines dead on one side, the footload will be very heavy.

(ii) *Landing.*—A circuit in either direction can safely be made irrespective of which engines have failed. While manœuvring with the undercarriage and flaps up maintain a speed of at least 130 knots. Aim to have the undercarriage locked down at the end of the downwind leg. The flaps should not be lowered until the final approach is commenced and it is certain that the airfield is within easy reach. The two live engines should be used within the limits of rudder control to regulate the rate of descent. Speed and power should gradually be reduced, aiming to cross the airfield boundary at the normal engine-assisted approach speed.

61. **Undercarriage emergency operation**

If the hydraulic system fails, the undercarriage can be lowered by compressed air from a special bottle or bottles, irrespective of the position of the undercarriage lever.

The flap selector should be neutral before using the emergency air system.

The knob (80) for working the air system is just forward of the engineer's panel. The undercarriage cannot be raised again by this method. Although it will lower by this method irrespective of the position of the normal selector, the latter must be selected DOWN for landing before operating the emergency air system, and left in the down position after landing ; otherwise any leakage of air pressure may cause the locks to be released and the undercarriage to collapse.

62. **Flaps emergency operation**

After lowering the undercarriage by turning on the emergency air cock, the flaps may be lowered by operating the flaps control, which admits the air pressure to the flaps system. The flaps can be raised again, but there may not be sufficient air pressure to lower the flaps a second time ; furthermore it may cause the header tank to burst. If it is absolutely necessary to raise the flaps by the emergency method extreme care must be taken to raise them slowly by stages. If the flaps are lowered by the emergency method before landing they must be left down after landing, owing to the likelihood of bursting the header tank, if they are raised.

63. **Flapless landings**

The initial approach should be made at 120 knots ; little power will be required to maintain this speed. The approach is flat with a nose-up attitude, but control remains satisfactory. Aim to cross the airfield boundary at :—

45,000 lb	60,000 lb.
100 knots	115 knots

The touchdown is straightforward and the aircraft can easily be brought to rest within 2,000 yards.

PART IV—EMERGENCIES

64. **Bomb jettisoning**

(i) Open the bomb doors and check visually that both are fully open. See para. 23.

(ii) Then jettison the containers first by the switch (15) on the right of the front panel.

(iii) Jettison the bombs by the handle (16) beside the container jettison switch.

(iv) Close the bomb doors.

65. **Parachute exits**

The hatch in the floor of the nose should be used by all members of the crew if time is available; originally it was released by a handle in the centre, lifted inwards and jettisoned, but when Mod. 1336 is incorporated the hatch is enlarged and is opened by a handle at the port side. It opens inwards and is secured by a clip which holds the hatch up on the starboard side. It can also be opened from outside the aircraft.

66. **Crash exits**

(i) On Mks. 1, 3 and 10 aircraft, three push-out panels are fitted in the roof (one above the pilot, one just forward of the rear spar, and one forward of the mid-upper turret) except when Mod. 977 (which moves the mid-upper turret forward) is incorporated, in which case the third panel is deleted.

(ii) On Mk. 7 aircraft there are two push-out panels in the roof, one above the pilot, and one just forward of the rear spar.

67. **Dinghy**

(i) A dinghy stowed in the starboard wing may be released and inflated :—

(a) from inside by pulling the release cord running along the fuselage roof aft of the rear spar.

(b) from outside by pulling the loop on the starboard side, rear of the tail plane leading edge.

(c) automatically by an immersion switch.

68. **Ditching**

If ditching is inevitable :—

(a) The undercarriage should be kept retracted, but the flaps should be lowered 30° to reduce the touchdown speed as much as possible.

(b) All crash exits should be opened (see para. 65).

(c) Safety harnesses should be kept tightly adjusted and locked, but R/T plugs should be disconnected.

PART IV—EMERGENCIES

(d) If available the engines should be used to help make the touchdown in a tail-down attitude at as low a forward speed as possible.

(e) Ditching should be made along the swell, or into wind if the swell is not steep.

69. **Engine fire-extinguishers**

Each engine is provided with a fire-extinguisher system. When Mod. 1221 is incorporated four fire warning lights (one for each engine) on the front panel indicate if there is fire in an engine and the pilot is thus warned to stop the engine, feather the propeller and then press the appropriate fire-extinguisher button (22). When Mods. 1067 and 1314 are incorporated the fire warning lights are mounted on the respective feathering push-buttons (19), and if a fire warning light comes on, pressing the button also operates the fire-extinguisher system. The pilot should, however, press the fire-extinguisher button as well. If the warning light is not on, pressing the feathering pushbutton will not operate the extinguisher. The fire-extinguishers are also operated automatically by a crash switch.

70. **Emergency equipment**

(i) *Hand fire-extinguishers*

One on the starboard side of the air-bomber's compartment.
One on the port side of the pilot's seat.
One on the starboard side forward of the front spar.
One on the starboard side aft of the mid-upper turret.
One on the port side of the rear turret.

(ii) *Signal pistol*

This is stowed on top of the front spar ; the firing position is in the roof forward of the stowed position. The cartridges are stowed in spring clips on the starboard side of fuselage just forward of front spar.

(iii) *Crash axes*

One on port side of fuselage forward of main entrance door.
One on starboard wall in front of rear spar.

(iv) *First-aid equipment*

An outfit is stowed on the starboard side of the fuselage aft of the main door.

A.P. 2062 A, C, F, H, K, L & M—P.N.
Pilot's Notes

PART V

ILLUSTRATIONS and LOCATION OF CONTROLS

Location of controls not illustrated.

Service	*Location*
Undercarriage warning horn test pushbutton.	Behind pilot's seat.
Cross feed cock.	On floor, just forward of the front spar.
Priming pump and cock (if fitted).	In each inboard nacelle.
Air intake heat control.	Left of pilot's seat.
Radiator shutter switches.	On starboard cockpit wall.
Ground/flight switch.	On starboard side aft of front spar.
Cockpit heat controls.	One each side of the fuselage just forward of the front spar.
	Two adjustable louvres in fuselage nose.
Oxygen master valve.	At forward end of oxygen crate.
Camera pushbutton control.	On starboard rail of cockpit.
Reconnaissance flare stowage.	On either side of fuselage forward of flare chute.
Flame floats or sea markers stowage.	On either side of fuselage adjacent to flare chute.
Nitrogen system cock (if fitted).	On starboard side aft of front spar.
Flying controls locking gear.	Stowed on starboard side aft of front spar.
Dinghy release.	Fuselage roof aft rear spar.
Engine fire warning lights (if fitted).	Front panel.

PART V—ILLUSTRATIONS

FUSES

Service	*Location*
Bomb gear fuses.	Inside junction box at forward end of bomb aimer's compartment.
Oil and radiator thermometer fuses.	Pilot's auxiliary fuse panel.
Radio fuses.	Navigation panel.
Mid-upper and underturret, call lights, and, on early aircraft, beam approach fuses.	Mid-turret position.
General services.	Main electrical control panel.

COCKPIT — FRONT VIEW

KEY TO *Fig.* 1

Cockpit—Front View

1. Instrument panel.
2. D.F. indicator.
3. Landing lamp switches.
4. Undercarriage indicator switch.
5. D.R. compass repeater.
6. D.R. compass deviation card holder.
7. Ignition switches (8).
8. Boost gauges (4).
9. R.p.m. indicators (4).
10. Booster coil master switch.
11. Fuel cut-off switches.
12. I.F.F. detonator buttons (inoperative).
13. I.F.F. switch.
14. Engine starter pushbuttons (4).
15. Bomb containers jettison button.
16. Bomb jettison control.
17. Vacuum change-over cock.
18. Oxygen regulator.
19. Feathering buttons (4).
20. Pneumatic and brakes pressure gauge.
21. Signalling switch box (identification lamps).
22. Engine fire-extinguisher pushbuttons (4).
23. Suction gauge.
24. No. 3 and No. 4 engine master cocks.
25. Supercharger gear change control panel.
26. Flaps position indicator.
27. Flaps position indicator switch (Mks. 1, 3 and 7).
28. Throttle levers (4).
29. R.p.m. control levers (4).
30. No. 1 and No. 2 engine master cocks.
31. Rudder pedal.
32. Boost control cut-out.
33. Signalling switchbox (recognition lights).
34. Identification lights colour selector switches.
35. D.R. compass switches.
36. Auto pilot steering lever (Mk. 4).
37. Compass deviation card holder.
38. Magnetic compass.
39. Undercarriage position indicator.
40. A.S.I. correction card holder.
41. Beam approach indicator (if fitted).
42. Watch holder.

COCKPIT—LEFT-HAND SIDE

KEY TO *Fig.* 2

Cockpit—Left-hand side.

43. Bomb doors control.
44. Navigation lights switch.
45. " D " switch.
46. Auto pilot main switch (Mk. 4).
47. Radio controller.
48. Seat raising lever.
49. Mixer box.
50. Beam approach control unit (if fitted).
51. Oxygen connection.
52. Pilot's call light.
53. Auto pilot attitude control (Mk. 4).
54. Auto pilot cock.
55. Auto pilot clutch.
56. Brakes lever.
57. Auto pilot pressure gauge.
58. Pilot's mic/tel socket.
59. Windscreen de-icing pump.
60. Flaps selector.
61. Aileron trimming tab control.
62. Elevator trimming tab control.
63. Rudder trimming tab control.
64. Undercarriage lever.
65. Undercarriage lever safety bolt.
66. Portable oxygen stowage.
67. Harness release lever.

KEY TO *Fig.* 3

Engineer's panel—Mks. 1 and 3 aircraft

68. Ammeter.
69. Oil pressure gauges (4).
70. Pressure-head heater switch.
71. Oil temperature gauges (4).
72. Coolant temperature gauges (4).
74. Fuel contents gauges (6).
75. Inspection lamp socket.
76. Fuel contents gauge switch.
77. Fuel tanks selector cock.
78. Electric fuel booster and transfer pump switches.
80. Emergency air control.
81. Oil dilution buttons (4).

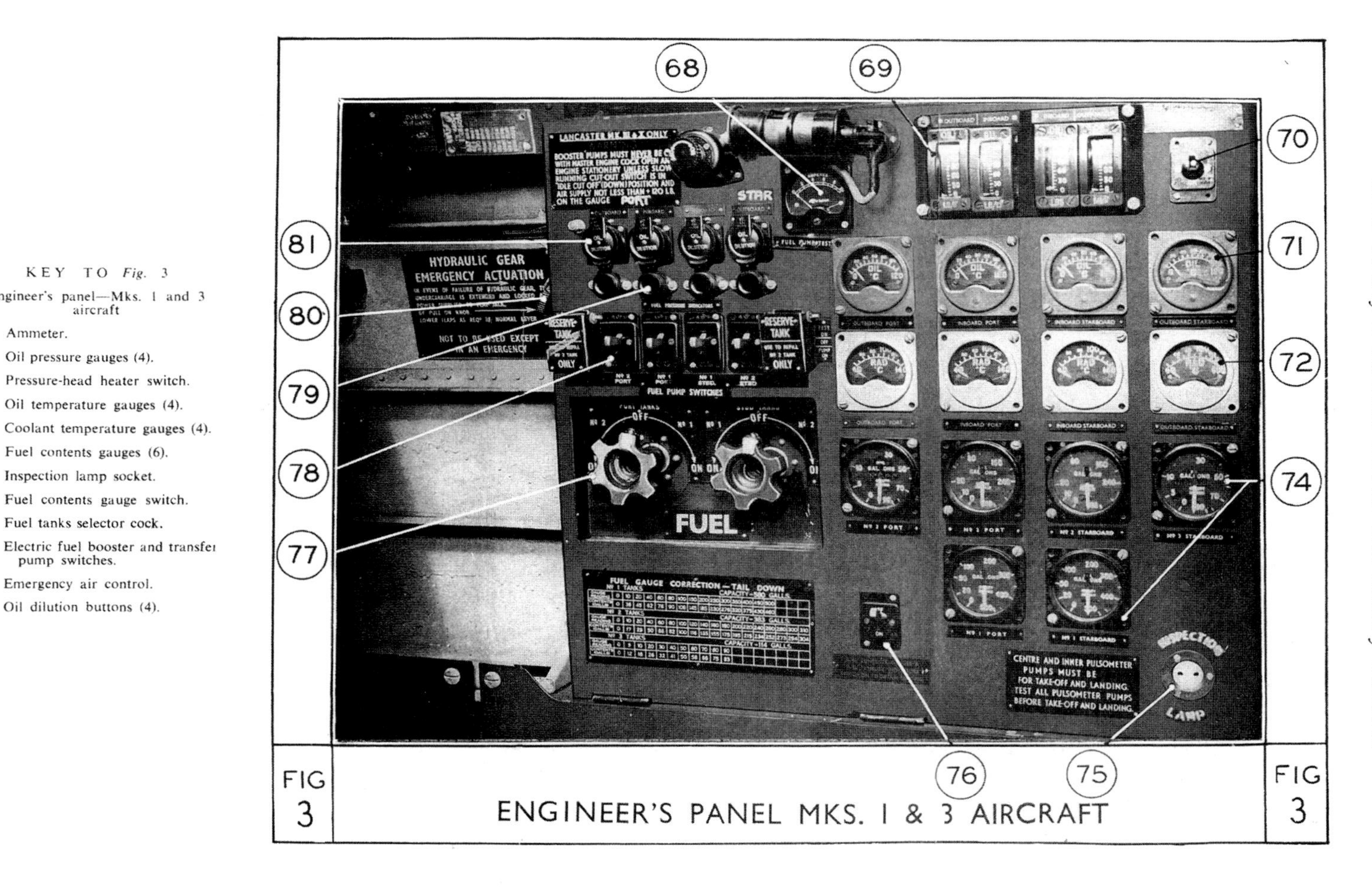

FIG 3 ENGINEER'S PANEL MKS. 1 & 3 AIRCRAFT FIG 3

FIG 4 ENGINEER'S PANEL MK. 10 AIRCRAFT FIG 4

KEY TO *Fig.* 4

Engineer's panel—Mk. 10 aircraft.

68. Ammeter.
69. Oil pressure gauges (4).
71. Oil temperature gauges (4).
72. Coolant temperature gauges (4).
73. Fuel pressure gauges (4).
74. Fuel contents gauges (6).
75. Inspection lamp socket.
77. Fuel tanks selector cock.
78. Electric fuel booster and transfer pump switches.
80. Emergency air control.
81. Oil dilution buttons (4).